Routledge Introductions to Development
Series Editors
John Bale and David W Smith

T0173795

Ecology and Development in the Third World

The improvement in living conditions of the inhabitants of the Third World involves both economic growth and ambient ecological conditions. Development of a country implies improvements in both spheres, yet these two factors are often in conflict.

Ecology and Development in the Third World introduces the links between problems of environmental degradation, economic development and population pressure in the Third World. Drawing on example from a wide range of countries, this book explains the machinery of environmental protection and stresses the importance of an integrated approach to ecodevelopment which applies technical, social, economic and political solutions to environmental problems.

Since publication of the first edition sustainable development has become a guiding principle in environmental management. Yet controversies still exist regarding what sustainable development is and how it can be achieved.

This second edition considers the many recent changes and events on the international stage, including the UN Conference at Rio and the new roles of the UN, the World Bank, national governments and people of the world in general in relation to managing our environment for the future. Thoroughly updated throughout, this edition includes much new material on global warming and ozone depletion, and environmental governance at various levels, as well as new case studies and illustrations.

Issues covered include: development and natural vegetation; environmental impact of land management; development of water resources; development and changing air quality; urban development and environmental modification; global concerns; concepts and mechanisms for global environmental management; and environmental problems and Third World development.

Covering all the major environmental issues and their management, the second edition of *Ecology and Development in the Third World* remains as concise and readable as the first edition.

Avijit Gupta is currently with the School of Geography at the University of Leeds.

In the same series

Avijit Gupta

Ecology and Development in the Third World

Second Edition

London and New York

To my father

First published 1988
by Routledge
2 Park Square, Milton Park, Abingdon, Oxon, OX14 4RN

Simultaneously published in the USA and Canada
by Routledge
270 Madison Ave, New York NY 10016

Transferred to Digital Printing 2005

Second edition first published 1998

© 1988, 1998 Avijit Gupta

Typeset in Times by
Pointing–Green Publishing Services, Chesham, Buckinghamshire

British Library Cataloguing in Publication Data
A catalogue record for this book is available from the British Library

Library of Congress Cataloging in Publication Data
Gupta, Avijit
Ecology and development in the Third World / Avijit Gupta. – 2nd ed.
      p.   cm.
Includes bibliographical references and index.
1. Sustainable development–Developing countries.
2. Environmental policy–Developing countries.   3. Developing
countries–Economic policy.   4. Natural resources–Developing countries.
5. Environmental protection–Developing countries.   I. Title.
HC59.72.E5G86   1998
363.7'056'091724–dc20   97–35347
                  CIP

ISBN 0–415–15192–9 (pbk)
ISBN 0–415–18631–5 (hbk)

Contents

Plates

Figures

Tables

Preface to the first edition

The improvement in the living conditions of the inhabitants of the Third World involves both economic growth and the ambient ecological conditions. These two factors need not be in conflict, and development of a country implies improvement in both spheres. This slender volume is an attempt to introduce the reader to the problems that are created when the ecological side of development is neglected. Attention is also drawn to the fact that we all share the same planet, that the environment is a wonderfully integrated system, and that any large-scale ecological misdemeanour may result in an ecodisaster for all of us.

I found my education and work experience in both the First and Third Worlds extremely useful in writing this book. I am also fortunate in having been taught for a few years in a university department which happens to bring together the two disciplines of geography and environmental engineering, an extremely uncommon combination, but very helpful in preventing one from looking at the environment from the narrow viewpoint of one's own interest.

I should like to acknowledge my debts to several individuals without whose help and encouragement this book probably would not have been finished. I am obliged to Irene Chee and Peh Mung Ngian for typing the manuscript and to Lee Li Kheng for drafting the diagrams with the usual speed and efficiency. The book has improved considerably from the criticisms on an earlier draft by A. Fraser Gupta, Mukul Asher and Richard Corlett. I should also like to thank the organizations and the individuals who gave me permission to reproduce their illustrations. Their names are included in the captions.

Lastly I should like to thank Anthea and Ella for their tolerance of my unsocial behaviour for the couple of months when the bulk of this book was written.

Preface to the second edition

The first edition of this book was written at the time when the Brundtland Report was coming out, and ozone depletion and global warming were topics being disputed by a number of people and organizations. A lot of changes have happened since then. Sustainable development has become a guiding principle in environmental management, although controversies are still very much in evidence regarding what sustainable development is and how it can be achieved. The United Nations Conference on Environment and Development was held in Rio de Janeiro in 1992. The United Nations, the World Bank, national governments and the people of the world in general are much more conscious of and knowledgeable about the state of the environment. Ozone depletion and global warming are not controversial any more. The discussions now concentrate on the extent and effects of global warming. The new version of the book includes the changes.

I have added two new chapters, one on global warming and stratospheric ozone depletion and the other on environmental governance at various levels. However, as far as possible I have kept the old format and the text, which have been popular, although new material has been added to all the chapters to bring them up to date and to stress the management side of the environmental issues. Three new boxed case studies have beeen added and bullets have been used to highlight major issues and examples. The new diagrams were, as in the previous edition, drawn by Lee Li Kheng. A. Fraser Gupta and Mukul Asher have read parts of the new draft. Anthea has continued to be extremely supportive.

1

Introduction

The countries of the Third World, with a few exceptions, possess certain characteristics in common. Most of these countries are in the tropics. Economically they are at a less developed state than the countries of western Europe or North America or Japan. They usually carry a large population, if not in absolute terms, at least relative to the opportunity offered by the environment. In general, the economy is based on agriculture, including herding, or on the extraction of timber or mineral products. Their progress is hampered by various disadvantages. These include:

- shortage of natural resources or technical expertise;
- inefficient transport networks;
- indebtedness to international financial organizations or developed nations;
- lack of power and influence in the international economic and political arrangements.

Within this general picture, however, there is a wide diversity. India, for example, has the second largest population in the world, over 935 million people in an area of 3,288 thousand km^2. It is a country of intense population pressure, low per capita income, self-sufficiency in food production, an established industrial base, and sufficient technical expertise to generate nuclear power and launch space satellites. Tanzania, with an area of about 950 thousand km^2 and nearly 30 million people, has an agricultural base and, in spite of the existing power and mineral resources, an underdeveloped industrial sector. Papua New Guinea, with a scattered population of about 4.3 million in a vast area of mountains, valleys, and coastal plains, is dependent on traditional agricultural practices and on extraction of forest timber or minerals such as gold and copper. Malaysia, a rapidly prospering nation of nearly 20 million people, is changing from a primarily agricultural country to one with more emphasis on industry and on the service sector.

The Third World countries are trying to improve the living conditions of their citizens. However, the steps taken to achieve this – the logging of timber, the extraction of mineral resources, the expansion and intensification of agriculture, the establishment of industries – may all occur simultaneously with a progressive deterioration of the environment. It is impossible to have development without some environmental degradation. But with careful management such degradation could be kept to a minimum. Such management requires:

- knowledge of the existing physical and socio-economic environment;
- identification of the environmental impacts of economic activities;
- enactment of laws and regulations which protect the environment;
- the political will to apply such laws and regulations.

These requirements are rarely fulfilled.

The tropical ecosystem is a fragile environment and, especially in relation to the fertility of the soil, is easily disturbed. However, environmental degradation has happened throughout history. A good example is the ruining of the soils of Mesopotamia several thousand years ago by the establishment of an irrigation system which brought salt up from the saline groundwater to the agricultural fields. The amount of degradation reached alarming proportions in the developed countries earlier this century. There, over the last forty years or so, the folly of development without an assessment of its environmental impact has been progressively realized, and laws have been enacted to prevent runaway ecological disasters from taking place. A century ago George Perkins Marsh (1898) said: 'Man has too long forgotten that the earth was given to him for usufruct alone, not for consumption, still less for profligate waste.' As the developing countries attempt to improve their economic conditions, they are also contributing to the degradation of the environment.

Improved economic conditions are crucial to the Third World, where they are needed to better the quality of life or, in some extreme cases, to prevent starvation. It is thus necessary to remove timber from the forests, extract minerals from the surface rock layers, expand farming into areas of unreliable rainfall or steep slopes, increase power generation and establish industries of various types. It is also necessary to review such projects before and after implementation so that the deterioration of the environment, if it cannot actually be prevented, can at least be controlled.

Environmental management currently happens at three levels. Certain large-scale problems such as anthropogenic climate change require co-operation between the countries of the world and supervision at the global level. Each country needs to safeguard its own environment by designing a set of national environmental policies and regulations which must be executed. An example of this would be preserving the national forest resource. Environmental management is equally important at the local level; it may involve insisting that an industrial establishment clean up its waste water before discharging it into the local stream. Environment is best preserved when management is efficient at all levels. Unfortunately, such exemplary management is not as common as it should be.

No country enjoys complete control over its own environment. This is mainly due to two factors: the global nature of certain types of degradation, such as the destruction of the stratospheric ozone layer; and the demand for its resources from outside its boundaries. It will not be possible to manage efficiently the environment of the developing countries short of working international agreements. However, this should not be an excuse for not immediately intensifying environmental management at local and regional levels.

Some progress has been made in recent years. Environmental awareness has certainly reached almost everywhere. It is perceived that environment is everybody's business, not just the business of the government at various levels (local, state, national) or a global organization such as the United Nations or the World Bank. What is yet to be achieved is translating this awareness into execution. But it is within the realm of possibility.

The purpose of this book is twofold. First, it provides an account of the nature of ecological degradation associated with development in the Third World; and second, it reviews the steps that could be taken to prevent or reduce such deprivation. The preventative steps are usually a collection of technical, social and economic measures.

2

Development and natural vegetation

The natural vegetation of the tropics and subtropics

Three large regions of rainforest occur in the Amazon Basin, Equatorial Africa, and south and south-east Asia (Figure 2.1). Away from these areas, where rainfall decreases and becomes seasonal, the rainforest is replaced by tropical monsoon forest, tropical grassland with trees, a poorer type of tropical grassland and finally by semidesert scrubland. The vegetation outside the rainforest has been greatly destroyed, and survives only in protected or relatively inaccessible areas. In the mountains, the tropical rainforests are replaced altitudinally by mountain rainforests, subalpine forests, alpine forests, and finally, on ranges that rise beyond 3,500 m, a treeless vegetation community. Mangroves in suitable places protect the coastal environment.

Traditionally the inhabitants of the rainforest have been hunters and gatherers or shifting cultivators. In the latter type of livelihood, patches of the forest are cleared by cutting and burning, and crops such as yam, cassava (also called tapioca and manioc), bananas, sweet potato and fruit trees are planted. In spite of the luxuriance of the forest the soils are usually infertile, and are kept productive only by the continuous decomposition of the fallen leaves and branches. Once cleared, the productive capacity of the land decreases in a few years, weeds start to establish themselves, and the cleared plots are abandoned for newer patches. The old clearings are soon under a secondary growth of vegetation, which starts to replenish the soil with nutrients, thus permitting recultivation of the plots after a gap of years.

Figure 2.1 Distribution of the rainforests of the world

With development, however, rainforests are perceived as a storehouse of resources: mainly timber and charcoal or firewood. Between such extractive activities and cultivation of both shifting and sedentary types, the rainforests of the world are shrinking at a rapid rate. Each year thousands of square kilometres of deforestation occur in Central and South America (Figure 2.2), South-east Asia and Africa. As Figure 2.2 shows, large areas of rainforest disappeared by the 1980s in Central America, a trend which still continues in many parts of the tropics. The situation for the other types of tropical vegetation is even grimmer; they are being rapidly destroyed in the search for new agricultural land and firewood.

The forests of the developed countries, on the other hand, are stable, with a total area of about 20 million km^2. The destruction of these forests happened earlier; at present the remnants are preserved and managed as reasonably stable forest resource bases. The amount of timber extracted is monitored and is balanced by new planting elsewhere in the forest. Pressure from agricultural expansion is also non-existent.

In the 1990s, the Food and Agricultural Organization (FAO) of the United Nations released the results of a survey (WRI, 1994). According to their estimates the current world average annual rate of forest loss is 0.8 per cent of the total forested area. For tropical forests only, the annual loss is about 154,000 km^2. The highest loss is from two regions: continental South-east Asia and Central America and Mexico, where the rate is double that of the world average. Half the loss of the tropical forests happened in six countries only: Brazil, Indonesia, Congo, Mexico, Bolivia and Venezuela. Even where the forest survives, it is likely to be fragmented and degraded. Such fragments and the forest edges remain particularly vulnerable.

The concept of biodiversity

Biodiversity is an all-embracing term used to refer to the range of variations in the biological world. It is a contraction of 'biological diversity'. In popular understanding, biodiversity indicates the number of species in a particular area, although technically it has been used to denote other kinds of variations too.

Tropical rainforests are extremely rich in species. Only a small fraction of the total number has been identified, and we do not know what else exists in the forest. A number of the plant species of the rainforest are used for various purposes, as listed later in this chapter. The biodiversity of the rainforests diminishes, with their destruction as species disappear. Rainforests are also home for a multitude of animals, birds and insects. Disappearance and fragmentation of the rainforest destroy their habitats, leading also to their

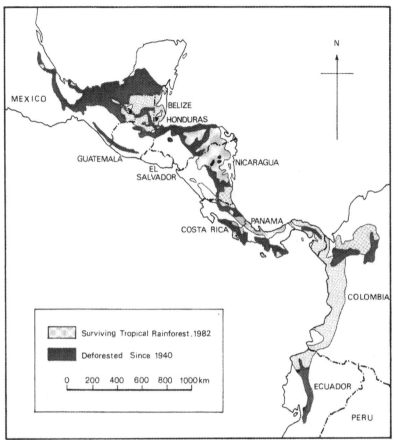

Lowland and lower montane tropical rainforests in Central America (Mid - 1983)

Country	Undegraded rainforest km²	Current rate of loss per year km²	Major threats
Nicaragua	27 000	1 000	cattle ranching
Guatemala	25 700	600	colonization, cattle ranching
Panama	21 500	500	cattle ranching, logging
Honduras	19 300	700	cattle ranching, colonization
Costa Rica	15 400	600	cattle ranching
Belize	9 750	32	colonization
Mexico	7 400	600	cattle ranching, colonization
El Salvador	0	0	(deforested)
TOTAL	126 050	4 032	

Figure 2.2 Destruction of rainforests in Central America, 1940–82

Source: J.D. Nations and D.I. Komer, 'Central America's tropical rainforests: positive steps for survival', *Ambio*, 1983, 12(5), 232–8

possible disappearance. Coral reefs form another habitat rich in biodiversity. Like the tropical rainforests, they are also under threat.

The demand for the resources of the forest

Apart from deforestation of the land due to timber extraction and agricultural expansion, tropical forests are also destroyed when mineral resources are discovered in the area, or when new highways and settlements are built. The extraction of wood is carried out at two different economic levels: tropical hardwood is extracted as an industrial raw material, and firewood is gathered by the poor section of the community.

Extraction of timber

Since the 1950s the extraction and export of tropical hardwood to Europe, Japan and the USA have increased tremendously. The light hardwoods of the Southeast Asian forests have been harvested at an alarming rate, especially in Indonesia, Malaysia and the Philippines. Huge timber concessions have been granted; for example, the Madang timber project in the Gogol Basin in Papua New Guinea, where a Japanese company extracted timber from a lowland rainforest for preparation of wood chips. Furthermore, tropical hardwoods provide a relatively homogeneous surface, and because of tree dimensions it is possible to produce large planks of uniformly high quality, which have numerous uses in the timber industry. The destruction of the forest is not restricted to the taking out of the valuable timber species. Collection of timber requires the construction of roads through the forest, and when large trees fall they destroy the vegetation in the vicinity. The tropical bamboo forests are under similar threat, bamboo being the raw material for the paper industry. Figure 2.3 shows the distance from which bamboo is collected for an Indian paper mill.

Collection of firewood

At the other end of the economic spectrum, forests, especially those in drier areas, are being reduced by the collection of fuelwood. Fuelwood and charcoal are the main sources of energy for the large number of poor of the developing countries. The use of firewood is extremely important in Nepal, India, China, Kenya, Zimbabwe, Brazil and Egypt, to name just a few examples. Every possible form of biomass is used: wood, twigs, crop residues, grass. If wood is difficult to procure, dried dung from domestic animals is used, thereby

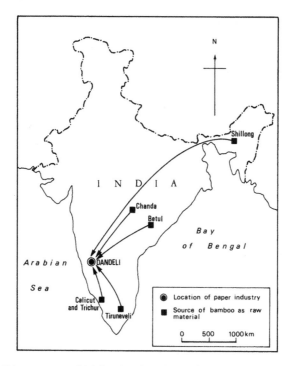

Figure 2.3 Distance over which bamboo is collected for a paper industry near the west coast of India

Source: A. Agarwal and S. Narain (eds), *The State of India's Environment, 1984–5*, New Delhi, Centre for Science and Environment

depriving the agricultural fields of animal manure. In the Himalayan Mountains, groups of women are forced to climb from the villages in the valleys to forests on the upper slopes to collect firewood. The time required disrupts family stability and shortens the time for what in other areas would be considered a normal day's household chores. Similar collection goes on in the African Highlands and in the desert fringes of the developing countries. As deforestation spreads, collecting firewood becomes more and more difficult over a very large part of the Third World (Figure 2.4). Firewood is in demand even in the cities, where it is hauled from distant forests.

Agricultural practices

Deforestation is also caused by agricultural practices. The increase in the number of shifting cultivators or the shrinking of the forest area available to

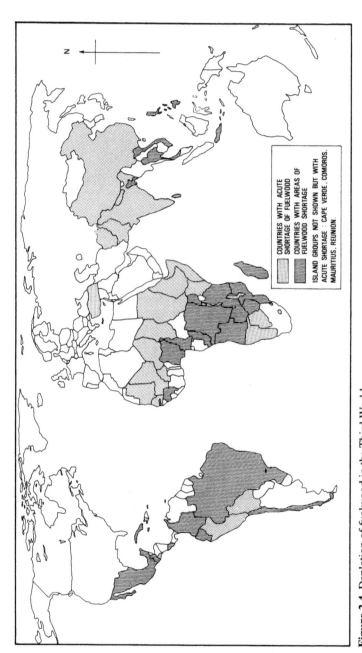

Figure 2.4 Depletion of fuelwood in the Third World

Note: Classification of countries based on World Resources Institute listing.

them necessitates a return to the old plots before the soil has had time to regenerate itself. Poor landless peasants tend to move into the edges of the forest in search of farming areas. Since the seventeenth century large areas of the tropical forests have been cleared for plantation in all the three continents concerned. For example, rubber and oil palm have replaced the rainforest on the lower slopes of Malaysia. The latest addition to the list are the cattle ranchers in the American rainforest. In Central America beef cattle are raised on pastures derived from the rainforest, a practice which is also found in the Amazon Basin. The lean meat from the grasslands cattle is used in the hamburgers, hot dogs and canned meat industries of the USA. The demand for lean meat is met at a much lower cost than that incurred by raising the beef in the traditional US cattle country. By 1981, one-third of Costa Rican land was under ranching, primarily for export of beef to the USA (Jackson, 1983). Similar cattle ranching has occurred in parts of Africa in response to the demands of the European Community.

Miscellaneous demands on the rainforest

The forest is also destroyed by other activities. Mining on a large scale, such as for the iron ores of Carajas in Brazil or the Bailadila area of Bastar district, India, destroys the flora and fauna over a large area. Highways like the Trans-Amazon Highway or the Trans-Sumatra Highway cut through virgin forest and bring in settlers, often as part of a government enterprise (as in the Indonesian *transmigrasi* project). Deforestation is the final result of all such demands on the land and resources of the tropical rainforest.

The effects of the deforestation

Two basic problems complicate the evaluation of the extent of deforestation. First, the data supplied by the governments of various countries are usually a few years out of date and the area of deforestation thereby understated. A better estimation is possible with remote sensing. On the island of Borneo in 1982–3, the extent of damage from the catastrophic fire, which apparently was triggered by burning for land clearance during a particularly dry season and which destroyed nearly 35,000 km^2 of forest, was measured from satellite photographs. Wth remote sensing it is possible to map not only the deforested areas, but also its recovery or replacement year by year. Even forests over the equatorial highlands, normally hidden from satellite sensors by clouds, can be mapped using radar. The second evaluation problem relates to the multiscale effects of deforestation. The local impacts such as increased soil erosion, decreased fertility and loss of flora and fauna are better known than the long-term and global effects of large-scale deforestation.

Soil erosion and depletion

The immediate effect of clearing the forest is accelerated soil erosion. It not only ravages the surface into a network of gullies, but also greatly increases the sediment load of the local streams, often choking them with bar formations. Less water infiltrates into the ground after forest clearance, especially if heavy equipment has been used, as such equipment compacts the surface. This increases surface runoff by decreasing infiltration of water into the subsurface. Frequently, the old logging roads become flood channels, red with sediment after a heavy rainstorm. The eroded sediment and water are quickly and simultaneously transported to a sediment-choked river, causing it to overflow. Thus an increase in flooding as a consequence of deforestation is to be expected. If steep slopes are cleared of forests, landslides start to occur frequently. Data from various parts of the world indicate that forest removal easily increases erosion by a hundred times or more by volume of material removed (Anderson and Spencer, 1991). As mentioned earlier, organic matter and other nutrients of the soil are quickly lost and not naturally replenished, leading to a rapid decline in crop yields. Overgrazing in drier areas destroys the scattered vegetation, leading to the same results of soil erosion and depletion.

Loss of local flora and fauna

Apart from timber, tropical forests supply oils, gums, rubber, fibres, dyes, tannin, resins and turpentine. They are also sources of many varieties of fruits and ornamental plants. Many of the tropical rainforest species have not been properly examined or even discovered. Some of those that are known are of tremendous importance. For example, tropical forests provide ingredients for drugs for leukaemia, Hodgkin's disease and contraceptive pills. They also supply strychnine, ipecacuenha, reserpine, curare, quinine and diosgenin – all invaluable raw materials for the pharmaceutical industry. Species from the tropical forests have been used in the hybridization of cereals which are responsible for the Green Revolution. This great gene pool of the tropical forests should be preserved. The forests are also home for a vast number of fauna, and the destruction of the rainforest is likely to result in the simultaneous elimination of a large number of such species. The depletion of the forest near its edges to extend cultivated areas leads to the movement of animals into croplands due to habitat destruction. This often results in unhappy confrontations between the farmers and the forest animals.

The macro-scale consequences

The global and long-term effects are not clearly understood yet, but certain consequences are feared. The loss of the forest decreases the moisture in the soil and, if large-scale, is also expected to lead to a reduction in the rainfall of the region. Together, these would lead to changes in the flow patterns of local rivers. The surface of the deforested land reflects a greater amount of solar energy back to the atmosphere. A macro-scale deforestation like that of the Amazon Basin may lead to changes in atmospheric heat flux and rainfall patterns on a global scale.

The biggest modification of the earth's climate is expected to come from an increase in carbon dioxide in the atmosphere following widespread deforestation. Such an increase on a global scale may provide enough carbon dioxide in the atmosphere to absorb sufficient thermal infrared radiation to cause a world-wide rise in temperature, a phenomenon labelled as global warming. Global warming is discussed in detail in Chapter 7, but it should be noted that saving the tropical vegetation is a significant step towards preventing such a change in the climate of the world.

Managing the ecodisaster

Management strategies pose the dilemma of either preserving the forests of the developing countries as an invaluable component of the physical environment or using them as a resource for short-term economic improvement. Not only does the destruction of forests provide firewood and land, albeit temporarily, for the poor, but the extraction of timber also supplies the governments concerned with revenue, often as foreign exchange. Myers (1986) is of the opinion that the pressure of international debt has served to promote cattle ranching in the Amazon Basin and to reduce logging restrictions in Ecuador, the Ivory Coast and Indonesia. Similarly, the growing of cash-producing non-food crops on farmlands has driven the subsistence farmers into more fragile environments in the Sahel countries. There is thus a strong conflict between proper management which would maintain the forest along with its utilization and the desire for an immediate trading or subsistence resource which could be used for national development.

Several techniques are being practised for better management of the resources of the tropical rainforests. Agroforestry where shade-tolerant crops are grown in conjunction with the forest trees is one example. Crops grown in various parts of the world in this fashion include cocoa, rubber, palms, cassava, maize and green legumes. The cultivation is done in small plots which can integrate with the forest and can support the poor peasantry living in or near the rainforest itself.

Cleared land near settlements but not in agricultural use can be reforested with species suitable for use as firewood, which is in great demand (Figure 2.4). The land remains vegetated, but there is enough opportunity to collect fuelwood for energy by people who would otherwise be compelled to go to areas of undisturbed vegetation in search of fuel. Such forestry practices, which go beyond the standard techniques of forest preservation and make a direct contribution to the well-being of the society, are known as social forestry. An example of this is given in case study A.

A forestry practice which is coming into vogue is to plant new forests with timber-producing trees on already degraded land in order to provide the area with vegetative cover and also to reduce the necessity to destroy virgin forest for timber. According to FAO estimates, nearly 438,000 km^2 of industrial and non-industrial forest plantations are in the tropics where most of the developing countries are located. Five countries (India, Indonesia, Brazil, Viet Nam and Thailand) cover 85 per cent of these plantations (WRI, 1994).

Large-scale projects, such as regional logging operations, the construction of highways or the transmigration of farmers in the rainforest, are carried out under government supervision and should be carefully monitored with a view to preventing the destructive impact of such projects on the fragile environment of the rainforest. Such projects are not always even economically successful, and in the long run may deprive the country of its natural resources. Imports of forest products are increasing in some developing countries. Until 1974, Nigeria was an exporter of forest products but since then it has been importing timber. The problem of large-scale forest destruction can only be solved under a global plan which deals with both forest maintenance and market demand. Such a plan is not really in place. In 1985 the World Resource Institute in conjunction with other institutions, including the World Conservation Union (IUCN), the World Bank, the United Nations Development Programme (UNDP) and the FAO, submitted a plan for arresting the destruction of the tropical forests (Table 2.1), and stressed that current trends in deforestation cannot be reversed solely by forestry but only with combined efforts from the agricultural and energy sectors, among others. Furthermore, such attempts have to involve local community groups, provincial and national governments and the development assistance agencies. A five-year attempt at improving the condition of the tropical forests according to this plan would require $8 billion

Table 2.1 Proposals for accelerated improvements in tropical forests

---

1 Improved fuelwood and agroforestry
2 Proper land management on upland watersheds
3 Improved forestry management for industrial use involving protection and management of natural forests, more intensive use of existing resources, accelerated industrial reforestation
4 Conservation of tropical forest ecosystem through better commitment by governments and development agencies and by strengthening appropriate government agencies
5 Strengthening the institutions for research, training and extension of services to the local community

---

Source: World Resources Institute (1985) *Tropical Forests: Call for Action*, Washington, D.C.: World Resources Institute

over a period of five years, a figure which indicates the scale of the problem. Another attempt was made at the 1992 Earth Summit meeting at Rio de Janeiro but the original version of the arrangement was diluted too much to have any significant impact. Another agreement reached at this meeting to save global biodiversity may, however, turn out to be useful in preserving the tropical forests.

It is also necessary to raise the awareness level of people away from the forested lands of the value of investment in their conservation. Important decisions about the future of the forests are taken in cities. Participation at various levels of the community is crucial to proper management practices which would allow the forest to be utilized without extensive environmental degradation. An example of conservation at the international agency level would be the Man and Biosphere Programme (MAB) of UNESCO which establishes biosphere reserves. These are multiple-use conservation areas that include both forests that are untouched and forests that are modified by human activities. The basic idea is to save a core of virgin forest by preserving around it buffer zones of somewhat degraded forests where research activities, visits from tourists and some agricultural practices by local inhabitants are permitted. The original forest is thereby preserved without confrontation, it is possible to carry out research, and controlled tourism at least partially funds the project. At a different level of conservation is the *Chipko* movement in India where local villagers have prevented the destruction of natural vegetation by hugging the trees as they were about to be logged by outside business concerns, thereby attracting public attention to the cause of preservation. The Kuna Indians on the offshore islands of Panama, indigenous people of the tropical rainforest, successfully maintained their traditional values in spite of outside pressure for development, and have established on part of their land a wildlife reserve with research facilities for visiting scientists.

Case study A

Social forestry in India

In India social forestry programmes have been started at the state government level in order to provide fuelwood, fodder, small timber, and minor forest produce to the rural people. Social forestry in India has three aspects:

- farm forestry, where the farmers are supplied with free or subsidized seedlings in order to encourage them to plant more trees on their land;
- community woodlots, where village communities are encouraged to plant trees on common lands to be equally shared;
- forestry woodlots, where trees are planted for the community by the government forestry department on public lands such as along the roads or banks of canals.

In general, it appears to be a successful project, but criticism has been levelled at the system in three areas. First, the programme, which is expected to use short-rotation trees as a cash crop, tends to benefit prosperous farmers more than the poorer section of the community; second, the community woodlot part of the programme is not progressing successfully; and third, the extensive use of eucalyptus as a cash crop undoubtedly provides fuelwood but by lowering the water table it depletes the soil of nutrients and in general degrades the area eco-logically. The advantage of social forestry is that it involves the society and government at various levels in preservation of the vegetative cover, which is the most successful way to attempt afforestation. In Gujarat a very successful afforestation programme has been carried out by pro-viding the schoolchildren with the seeds of the subabul tree and encouraging them to look after their trees, which in turn supply the community with fuel and fodder. The effect of social forestry programmes in India has been discussed in detail by Agarwal and Narain (1986).

Case study B

Changes in the Amazon rainforest

The 6 million km^2 Amazon Basin occupies over a third of South America. Most of it is in Brazil, but the headwaters of the Amazon also drain large areas of the Andean states of Bolivia, Peru, Ecuador, Colombia and Venezuela. In general, most of the Amazon Basin is under high rainforest (Figure B.1) on acid soils from which nutrients have been washed out. Unlike other rainforest areas, the soils of the Amazon forest are low in nutrients even where tree cover exists. The forest receives its nutrients through a network of fine roots at the soil surface, which is mixed with organic material. The roots absorb nutrients directly from the decomposed litter, so nutrients do not get into the soil in any significant quantity. The traditional inhabitants have been various small groups of Amazonian Indians such as the Yanomami, who live off the forest without creating any significant environmental stress.

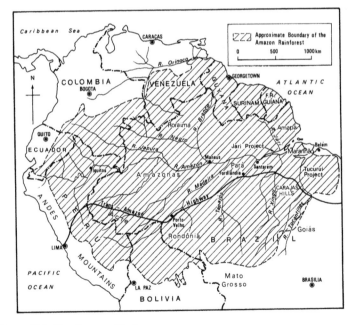

Figure B.1 The Amazon Basin

Case study B (*continued*)

To Brazil, trying to develop its resources, the sparsely populated forested vastness of the Amazon Basin is a great temptation. The forest contains large amounts of commercial timber and other forest products. It was naturally growing rubber (the plant *Hevea braziliensis*) that prompted the first large-scale alteration of the Amazon forest in the late nineteenth century. The cleared land attracts migratory peasants. The mineral deposits of the area are considerable, including large deposits of iron and manganese. The iron reserves of Carajas, south of Belem, have attracted large-scale mining. Gold and tin mining is also important, often leading to various kinds of environmental degradation, such as the spillage into the rivers of mercury used in gold processing. Serious confrontations between miners and local inhabitants occur. Forest-based industries have been set up, and recently large tracts have been cleared for cattle ranching. The Brazilian government, poor migrating farmers and large business organizations, including multinationals, have been involved in the development of the Amazon Basin.

Very few of the projects have been successful, and none has been implemented without grave environmental consequences. Two early attempts to develop the interior of the Amazon – the gathering of natural rubber about a hundred years ago, and Henry Ford's attempt to establish a rubber plantation between the mid-1920s and 1945 – both failed. Examples of recent large-scale projects include a number of foreign involvements, such as the US shipping executive Daniel Ludwig's timber, wood pulp and agricultural project at Jari over an area of 16,000 km^2. Such projects, often with government support at least at the beginning, have not always been successful, but have usually destroyed a large part of the rainforest, turned the land barren and increased erosion.

The building of the Trans-Amazon Highway and a connecting road network leading from the east into the Amazon Basin has been followed by the migration of poor farmers into Rondônia and Acre (Figure B.1). This in general has not been a successful migration, due to the difficulty in cultivating the soils of the Amazon tropical frest, and often has given rise to a sequence of forest destruction, soil depletion and erosion, and a further migration by the disappointed peasantry. Large areas have been converted from forest to pasture for raising beef cattle. Although most

Case study B (*continued*)

of the basin is a great plain, innumerable streams have cut deep furrows into the surface, causing rapid erosion of the soil.

A number of hydroelectric projects are in operation in the basin or are being constructed. Reservoirs associated with these projects often submerge very large areas of the rainforest. For example, the Tucuruí Dam on the Tocatins River in the eastern Amazon Basin has flooded 2,000 km^2 of forests (Salati *et al.*, 1990).

The results of such alterations are perceived at two levels: regional and global. The cleared areas of the Amazon rainforest suffer from soil infertility, erosion and a possible increase in flooding. The global effects have the potential to be extremely serious.

The Amazon carries about 20 per cent of the world's total river discharge and about 10^9 t of sediment to the sea each year. Large-scale deforestation in the basin will ultimately alter the Amazon's discharge regime and sediment transport pattern. Changes at that level may have grave environmental consequences, and probable significant climatic changes have been predicted.

The destruction of the huge vegetation cover leads to a build-up of carbon dioxide in the atmosphere. This would block the escape of terrestrial heat and raise the surface temperature. The scale of deforestation in the Amazon Basin has the potential to accelerate such global warming, discussed in Chapter 7. The third major aspect of the global disaster would be the deplorable destruction of the vast number of species (both flora and fauna) that the Amazon rainforest currently holds, a huge loss of biodiversity.

Key ideas

1 Only a part of the original natural vegetation has survived in the Third World.
2 The three major areas of surviving tropical rainforests are the Amazon Basin, Equatorial Africa and South and South-east Asia.
3 The rainforests should be perceived as storehouses of resources.
4 The tropical vegetation is disappearing rapidly to meet the demand for timber, firewood and new agricultural land.
5 Fuelwood is an important source of energy in the developing countries.

6 Forests have been replaced by plantations or cattle ranches in various parts of the tropics.
7 The effect of the deforestation, depending on the scale, could be local or global.
8 Deforestation leads to the loss of biodiversity, which is a serious environmental degradation.
9 The management of the tropical forests should be carried out simultaneously at various levels: international, national and local.

The environmental impact of land development

The background

Land development, especially extension of farming areas, happens mostly in response to a rise in population or a rising demand for a particular crop (Plate 3.1). Pressure on land is created when an area is expected to support more than the optimal number of people. It is not only the population size that creates pressure on land but also the crowding of people in the more advantageous geographical areas within a country and the migration of less-fortunate people to marginal areas in search of agricultural land. Even countries with an overall low population may end up with high density in certain regions. For example, the concentration of the rural population of Venezuela is along the Andean foothills, not on the lower plains or the southern uplands. The population in western Malaysia is distributed along the two coastal plains, avoiding the central mountains. Pressure on land is also created by external interests, as is the case in the clearing of the Central American forest for raising hamburger-destined cattle. In addition, large-scale agricultural developments, the establishment of valley-bottom water reservoirs, the spread of industries or of urban settlements, all can create pressure on land by occupying the good agricultural land, and forcing the peasantry to migrate to other areas.

Whatever the underlying causes, agricultural expansion involves the settling and farming of marginal lands: areas which are difficult to cultivate and which would not be the first choice of most rural communities. The marginality is displayed in steep slopes, low rainfall, poor soils and so on. It is a fragile environment which is very easily disturbed. Farming of such land results in

Plate 3.1 Rice grown on steep slopes, north Sumatra

erosion and exhaustion of the soil, accelerated slope failures, salinization and desertification. Interestingly, the problems created by the misuse of land and their solutions are not necessarily all physical. In many cases a socio-economic appendage to the physical solution is needed. The application of technological solutions to ecological problems has to take into account the historical or cultural background of the people concerned, in order to persuade them to utilize the land properly.

El Salvador is not a unique case. It is a typical example of the situation where the degraded landscape cannot be recovered solely by the technical solutions of reforestation, conservation measures, river control and so on. Further development of an export-orientated economy could very well worsen the problem by forcing even greater use of land for large plantations. Part of the solution has to come from a modification of agricultural practices and the land tenure system. Population control, so important in preventing environmental degradation, has very little chance of being effective unless the poor farmers are assured of a better socio-economic existence. The profits from commercial agriculture, which improve the gross national product (GNP) of a country, are seldom used to improve the environment or the fate of subsistence farmers. A landless tenant farmer or a subsistence farmer, whose family is perpetually under the threat of malnutrition, cannot be expected to be always concerned

Case study C

Land degradation in El Salvador

El Salvador is 20,000 km^2 of the volcano-studded mountainous backbone of Central America sloping steeply down to a narrow coastal plain by the Pacific Ocean. The population of 775,000 at the beginning of the century is now approaching 6 million. Such a dense population exerts a heavy ecological pressure in this predominantly agricultural country. At one time, about 90 per cent of El Salvador, especially the steep mountainous country, was under tropical deciduous forest. Centuries of grazing, mining, charcoal burning, the spread of plantations and subsistence agriculture have destroyed the forest. Not surprisingly, about three-quarters of El Salvador, especially the mountain slopes, are ravaged by accelerated erosion. Soil erosion on slopes contributes to filling the river channels with the eroded material as it travels downslope (Eckholm, 1978).

The degradation of the environment is, however, not entirely due to the size of El Salvador's agricultural population. The pattern of land ownership contributed greatly to the ecological disaster. About half of El Salvador has for a long time consisted of large estates called *latifundias* occupying the best lands (coastal plains, valley flats, basin floors and middle slopes of volcanoes), growing coffee, sugarcane and cotton for export. A very small part of El Salvador's good agricultural land, on the other hand, is fragmented into a large number of small plots of subsistence farmers; many small farmers are thus forced onto marginal lands on steep slopes and low-fertility soils; in order to survive, they have to cultivate their plots without proper fallowing practices, starting a vicious circle of progressively degraded land.

---

about the degradation of the environment. A government that is interested in protecting the national environment must communicate successfully the notion of impending ecological disaster across a wide spectrum of people, from the farmers to the politicians and bureaucrats.

Agricultural development: the general case

Agriculture in the developing countries has supported a very large number of people for a very long time (Plate 3.2). The traditional agricultural systems include shifting cultivation in the forests, sedentary farming in fertile regions

Plate 3.2 Rice and coconut farming, Bali

(flood plains, coastal plains, volcanic slopes), irrigated farming in subhumid areas, and grazing livestock in drier or uncultivable tracts. With the rise of population in recent years, traditional agricultural techniques have become insufficient to meet the demand for food in many places. As a result, various modifications of the traditional systems have taken place, not always without ecological problems.

The surviving tropical rainforests of South and Central America, Central Africa, and South and South-east Asia still cover a large area. The luxuriance of these forests perpetuates the myth that the tropical lands are indigenously fertile. Due to the efficient leaching of the soil, the tropical lands are naturally productive only if any of the following situations occur:

- the regular renewal of the fertile constituents of the soils by decomposing leaves and branches on the forest floor;
- the repeated flood alluviation on riverine plains;
- the formation of virgin soils on fresh volcanic slopes.

The traditional agriculture in the rainforest has been shifting cultivation practised by a small number of forest-dwellers: a patch of land is cleared and burned and a combination of crops is cultivated for a few years. The plot is then abandoned following the drop in fertility and the proliferation of weeds. Usually the system implies a return to the abandoned plots in a number of years after renewed forest growth, and therefore this system is able to support only a limited number of practitioners. Where such constraints can be met, the system is ecologically sound and land degradation is limited. An increase in population or a decrease of the forest area requires a return to old plots before the soil has had time to recover, with disastrous consequences in falling yields, general malnutrition and increased soil erosion. The soils of the tropical rainforest can be cultivated continuously only with massive doses of fertilizers. In addition, due to the intensity of tropical rainfall, any land not under natural vegetation or crops is rapidly eroded by surface runoff, rills and gullies. Many of the Third World countries have a colonial past. History is full of the failures of the colonial masters who tried to impose a system of continuous cultivation on the soils of the cleared rainforest.

Even in areas that are naturally fertile or where animal manure has been carefully used, as in the highlands of East Africa (Stáhl, 1993), the population density is causing too much pressure on the environment, leading to soil erosion and increased flooding of rivers. Spectacular effects of erosion are seen in Madagascar where heavy rains fall on steep slopes from which vegetation has been cleared. Even the ecologically stable system of the production of irrigated rice in the wet fields of Asia is under pressure, as in the Ganga Valley of India or on the island of Java in Indonesia. Nearly all the good land in the world is already in cultivation, which highlights the problem, especially with a rising global population (Kendall and Pimentel, 1994).

The solution to this problem arrived in the late 1960s and early 1970s in the form of an agricultural practice known as the Green Revolution. The Green Revolution is the result of decades of painstaking genetic research on various kinds of crops in several outstanding research projects, such as Dr Norman Borlaug's wheat-breeding programme in Mexico and the work in the International Rice Research Institute in the Philippines. In essence, the Green Revolution requires the use of improved high-yield varieties of grains like wheat or rice grown with large inputs at appropriat times of fertilizer, water and pesticide. The doubling or more of wheat and rice yields has enabled countries like Mexico, India, Pakistan, the Philippines and Turkey to meet the demands of an increasing population. There has been a phenomenal increase in world grain output from the 1960s onwards, and a very large part of it is due to the Green Revolution.

Certain environmental problems, however, may arise from this type of development:

- it is easier for the more prosperous farmers to take advantage of the Green Revolution, thereby increasing the disparity of wealth among the members of the community, which may or may not have a physical manifestation;
- the fertilizers and pesticides wash off the fields to pollute rivers or lakes, or leach into the subsurface to lower the quality of the groundwater;
- both fertilizer production and irrigation (which often uses diesel pumps for lifting water to the fields) require energy, and the rise in the price of oil and natural gas since 1973 has made energy a much more expensive commodity;
- there may be biological complications, such as the genetic resistance to pesticides developed by various insects or a possible reduction of genetic diversity in crops.

But, on the whole, the Green Revolution has been successful even if it has only bought time for the planners of population control.

Agricultural development: the marginal lands

The degradation of the environment as a result of agricultural expansion is, as expected, strongest in marginal lands. The expansion is connected with large increases in population and inequitable land tenure systems. There are three major types of marginal lands where such expansion takes place:

- areas where agriculture is possible with irrigation;
- drier areas with limited potential for crop growing but where grazing is possible;
- areas of steep slopes with a very high erosion potential.

Agricultural development with irrigation in sub-humid areas

Land has been successfully cultivated under irrigation for thousands of years, starting with the manual or animal-driven lifting of water from rivers and ponds to the fields. The general practice of putting a barrier across a river to pond water and then redistributing the stored amount to agricultural fields via a network of canals is found throughout history, especially in sub-humid, semi-desert areas like the Tigris–Euphrates Valley or the coastal desert area of Peru. Both in earlier times and in recent years, a network of earth canals and waterlogged fields have led to seepage underground and a subsequent rise of the water table in semi-arid regions. When saline groundwater, as is found in these areas, rises to within a few metres of the surface, capillary action brings it right up to the surface. The water evaporates, leaving behind a layer of salt which makes farming impossible. Where this phenomenon is advanced, entire areas are covered with glistening white salt. In order to prevent such a condition, irrigation in sub-humid areas should be planned concomitantly with provisions for better drainage so that waterlogging of the fields does not occur, and the canals should be lined with bricks or concrete.

The list of countries affected by salinity caused by the spread of irrigation water is extensive. In the developing world it includes the drier parts of India, Pakistan, Iraq, Iran, Syria, Jordan, other countries of West Asia, Peru, Argentina, north-east Brazil and Mexico. Sometimes salinity is exacerbated by the human hand, as in the Colorado delta. The water that flows down the Colorado River from the USA to Mexico is highly saline, due not only to the local geology, but also to the increased diversion of water for agricultural use and its subsequent return to the river. The evaporative loss of part of the irrigated water in transit makes the remainder even more saline. An agreement was reached in 1973, by which the USA agreed to build a desalination plant to lower the salt content of the water draining to Mexico. The spread of irrigation in sub-humid areas requires simultaneous improvement in drainage conditions to prevent the accumulation of water in the fields. India is currently involved in the huge scheme of the Rajasthan Canal to turn a vast area of the Thar desert and its fringes into productive agricultural land, and the effect of the drainage of the irrigated areas has to be carefully monitored.

The drier lands

About a third of the world is classified as desert of various types and degree. Around the fringes of the major deserts, such as the Sahara or Takla Makan, there is usually a transitional belt which will support nomadic herdsmen in years of good rainfall. At other times, the fringes are too inhospitable even for that. The Sahel of Africa is such a fringe, where years of drought bring starvation and famine to the people of Mauritania, Senegal, Mali, Burkina Faso, Niger and Chad. Since the 1970s this has been disastrous enough to attract the attention of the world periodically (Figure 3.1). In recent years there have been two periods of disastrous drought in Africa: 1968–73 and 1982–4. During the first event, thousands of people perished along with a very high percentage of livestock. The estimate of 100,000 death, publicized about the time of the drought, however, has been queried later (Olsson, 1993).

Around the Sahelian type of desert fringe lies a semi-arid environment, which is climatologically less hostile. Such an environment is found in part of all the west African countries listed above, and also in Sudan, Somalia,

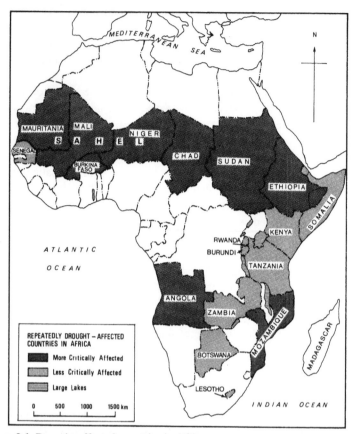

Figure 3.1 Drought-affected countries in Africa

Source: M.H. Glantz and R.W. Katz, 'Drought as a constraint to development in sub-Saharan Africa', *Ambio*. 1985, 14(6), 334–9

Ethiopia, Kenya, Tanzania and in parts of southern Africa, especially Botswana. In Central and West Asia it includes the countries from Israel to Pakistan, the Thar of India; and in South America, parts of Argentina and northern Chile. Much of this area has been supporting more than the optimal number of people and grazing animals. This has resulted in the destruction of natural vegetation, overgrazing, burning and careless farming, leading to the erosion of the landscape, windblown sand, salinization, and finally to a dry desert-like land. This process is called desertification.

Desertification is helped by the strong rainfall variability in these areas, and the occasional pattern of dry years tending to run together. A population of humans and grazing animals that is not a pressure on land during wet years becomes insupportable during a dry spell, often propagated by the short-term climate change known as the El Niño-Southern Oscillation. The desertifiation of the Sahel area also has been accelerated by overgrazing and the emphasis in colonial times on cash crops such as cotton, groundnuts, and soybeans replacing food crops. The control of human and animal population, the restriction of human and animal migration to such areas and judicious farming practices including controlled grazing may prevent desertifiation. Recovery of such degraded areas might be possible with large-scale afforestation as has been attempted in Algeria, carefully planned irrigation projects, and use of new cropping systems that maximize the vegetation cover on land. It should be remembered that the areas currently under desertification have supported a number of people for thousands of years, although the climate may have been kinder in the past.

The areas of steep slopes

Agricultural expansion on hillslopes usually ends in intensive erosion of the slope materials, and transport of the sediment downslope to the streams at the bottom of the valleys. There the sediment raises the river beds, chokes the channels and increases the potential for flooding. The effect of deforestation followed by agricultural expansion is easily seen when the sediment loads of streams in forested catchment areas are compared with those in cultivated areas. The erosion occurs in many ways: slopewash, gullies and mass movements such as debris slides or debris flow. Even the standard soil conservation techniques of cultivating on the terraces on hillslopes or contour fencing might not be sufficient to prevent slope degradation. Only large-scale afforestation of the slopes seems to be effective against erosion.

The speed of degradation is usually high, but certain areas like the Himalayas, the Andes or parts of the East African Highlands are being eroded at an even faster rate because of the local geology, relief and the nature of intense rainfall over the slopes. Java's rivers carry a very high volume of sediment derived from erosion-prone mountain slopes. The highlands of Ethiopia are being heavily eroded, as demonstrated by the sediment load of the Blue Nile. The combination of bare mountain terrain and population pressure (Ethiopia is the third most populous country in Africa) results in a landscape of bare subsistence and famine, desolation and impending disaster, with sediment-laden rivers shifting their courses at the bottom of the hillslopes. A similar pattern of land degradation primarily arising out of population pressure is found on the Andean slopes of Bolivia, Peru and Venezuela. Studies in Colombia show an acceleration of slope erosion, landslides and valley-bottom sedimentation in similar fashion.

Perhaps the best studied and most spectacular examples come from the Himalayas, where a combination of local geology, steep slopes on high mountains and intense monsoon rainfall expose the disastrous folly of deforestation, overgrazing and cultivation of steep marginal lands. The catchment of the Kosi River, draining out of the high Eastern Himalayas and one of the most spectacular river basins in the world, has also been described as one of the most eroded ones in the world. Similarly, misuse of land in the upper Indus Valley of Pakistan releases enormous quantities of sediment which has effectively shortened the life of the huge Mangla and Tarbela reservoirs. This reduction in the life expectancy of reservoirs due to bad land-use practices has been reported from many countries including some in the developed world. Both agricultural expansion and grazing have led to deforestation and subsequent erosion of the Himalayan slopes.

Other types of land degradation

Other types of land degradation problems locally may assume disastrous proportions. The formation of badlands in semi-arid areas via the development of gully networks or the despoiling of the landscape by mining are two examples. A very large part of the Chambal Valley of central India is being rapidly eroded by a spreading network of gullies and streams known locally as ravines.

Mining practices degrade the landscape by:

• laying barren the land and opening the huge chasms of opencast mining;
• polluting the water from drainage of mined materials;
• denuding forests both through direct deforestation to extract minerals and by releasing pollutants into the air that cause acid rain;
• choking the landscape with a layer of dust that settles on the agricultural fields.

In the USA, for example, a reclamation plan has to be filed and approved by the government before mining for coal can begin. Such practices should be followed in less-developed countries in spite of the lure of economic development based on the extraction of coal and minerals.

The rivers and coasts of the developing countries exhibit too many cases of excessive sediment being released into the waters. The results may be limited because of the dilution involved, but one wonders about the effect on neighbouring mangrove assemblages and coral reefs. Harmful chemicals also ride piggyback on sediment grains and, as a result, the mouths of a number of rivers and estuaries are seriously polluted. The development of the land should be planned so that the impact of deforestation, agricultural expansion and mining are minimized. This can be achieved only by a combination of technical, social and economic solutions. Crop management, soil conservation measures and simultaneous establishment of irrigation and drainage have a higher rate of success when they arrive with population control and land reforms.

Case study D

The Yallahs Valley, eastern Jamaica: an example of the utilization of marginal lands

Eastern Jamaica has a deeply dissected mountainous centre surrounded by narrow coastal plains. The 37 km long Yallahs River drains 163 km^2 of the southern slopes of the central Blue Mountains (Plate D.1) into a basin with the impressive and persistent local relief of 450–600 m, with steep slopes at angles of 20°–30°. Extensive landslides and faulting help to maintain this landscape. The annual rainfall increases from 1,500 mm near the mouth of the Yallahs River on the southern coast to more than 2,500 mm in the north over the upper Yallahs basin. There is a seasonal component in the rainfall, December to April being the dry season. Some of the rainfall arrives in the form of violent and intense showers, at times from tropical storms that occasionally reach hurricane force when rainfall of nearly 250 mm a day is not rare. The soils are thin, full of pebbles, not particularly fertile and heavily eroded by mass movement, slope wash and gullies.

An export-orientated agriculture resulting in the cultivation of sugarcane, coffee, tobacco, cotton and cinchona has replaced most of the natural luxuriant forest vegetation over the last 400 years. This started as plantation agriculture in large estates using slave labour, which mainly produced sugarcane and, higher up in the mountains, coffee or cinchona. A large number of plantations were abandoned in the mid-nineteenth century due to the emancipation of slaves in 1838 and the abolition of

Case study D (*continued*)

Plate D.1 The Yallahs River, Jamaica

of coffee plantations were abandoned about the same time. By the second half of the nineteenth century, slopes that had been cleared were either utilized for plantation or subsistence agriculture, or were allowed to develop a secondary growth of grasslands and thorn trees, locally known as the *ruinate* lands. Another spell of land clearing happened towards the end of the nineteenth century when the cultivation of bananas both in small holdings and on plantations became commercially viable in Jamaica.

The spread of agriculture due to population pressure was remarkable, particularly considering the minute size of the holdings, the infertility of the soils and the steepness of the slopes where even terracing is not carried out and a fence of brushwood tied together at the downslope end of the plots functions as the soil-protecting device instead. In the Yallahs valley, human endeavour arising out of population pressure has speeded up the already rapid process of erosion.

Case study D (*Continued*)

At present, most of the Yallahs valley slopes are under *ruinate* bush, strip cultivation or pasture with some forests, and afforestation attempts. The general picture is that of smallholdings with vegetables, fruit trees and pasture with some coffee on the upper slopes. The Yallahs Valley Land Authority is responsible for planning and monitoring the use of the land, but it is difficult to utilize a landscape such as this (Plate D.2) without creating extreme erosion on slopes and floodprone rivers at the bottom of the valley. Along a hill road after the heavy rainfall from a tropical storm an average of ten landslides would be found in each kilometre. The recent expansion of new coffee plantations at the cost of forested slopes has not helped. It is interesting to note that the land capability map of Jamaica lists almost the entire basin as either not or only marginally suitable for cultivation, and recommends that it remains under natural vegetation, or tree crops or pasture.

Plate D.2 The eroded hillslopes in the Yallahs Basin

Key ideas

1 Pressure on land is created by many factors: high population density, non-uniform population distribution, skewed distribution of the size of land holdings, and large-scale development projects.
2 The Green Revolution has been very successful in increasing crop production in some countries.
3 Agricultural expansion to marginal areas has to cope with two types of problems: shortage of water and erosion of land.
4 Extension of irrigation without drainage into dry areas may result in salinization of the land.
5 Extreme care should be taken in utilizing mountain slopes.
6 The problems of land degradation that arise from agricultural expansion require technical, social and economic solutions – all three.

4

Development of water resources

An introduction to hydrology

Nature manages her water resources extremely efficiently with techniques for self-purification and for managing on a finite budget by recirculation of moisture. The recirculation technique is known as the hydrologic cycle, which is best explained diagrammatically (Figure 4.1). About 97 per cent of Earth's water is stored in the oceans in a saline form. Almost the entire amount of fresh water is locked frozen in Antarctica and Greenland. A significant amount of the remainder lies at a considerable depth in the subsurface. The water that is normally available to us comes from the atmosphere, the land surface and the shallow subsurface; and constitutes an extremely small part of the total water inventory. Technologically it is feasible to acquire, process and utilize water from the oceans, ice sheets and deep underground, but the cost makes such efforts uneconomic. One of the favourite expressions in hydrology is that people expect water to be cheaper than dirt.

Part of the precipitation is interrupted by the vegetation, and is either evaporated back to the atmosphere or runs down the plant stems or trunks to reach the ground. The water that reaches the ground surface enters into the soil (infiltration), or fills small (1–2 cm deep) surface depressions and, if the rain is intense, starts to flow on the surface (surface runoff) to rills and gullies, and ultimately to larger streams. The water that has infiltrated into the soil fills the smaller pores in it, and drains downwards by gravity (gravity drainage) via the larger pores to the groundwater below. Most of the water stored in the soil (soil moisture) is available to plants via their root systems. The groundwater table is raised by gravity drainage, which accelerates the lateral flow of the

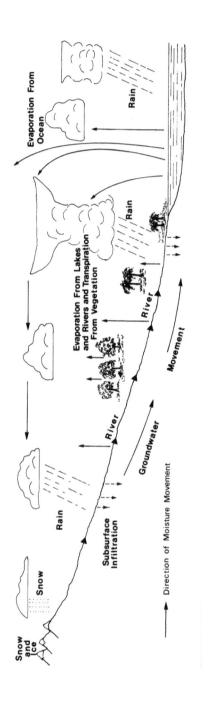

Figure 4.1 The hydrological cycle and water storage

Source: Figures in percentages are taken from R.G. Barry, 'The world hydrologic cycle', in R.J. Chorley (ed.), *Water, Earth and Man*, London, Methuen, 1969

subsurface water to wells and streams. The amount of water which will go via the different paths of this system is determined by local conditions and varies seasonally. Any development that alters the local environment alters the paths, and thereby the available amount and quality, of the water supply. Clearing land and urbanization increases both surface runoff and the potential for flooding of nearby streams. Afforestation, on the other hand, increases both interception and infiltration, and is expected to reduce floods and increase groundwater flow in the subsurface. Any water management scheme redistributes the amount of water along the various paths of the system, some of the results of which could be undesirable (Dunne and Leopold, 1978).

The demand for water

The story of the large-scale management of water goes back at least 6,000 years. Structures for storing water behind dams and for distributing stored water have been associated with ancient civilizations in various parts of the world: Egypt, western Asia, India and China. Some rivers, such as the Nile or the Tigris-Euphrates, have long histories of water utilization. At present, demand for water has risen to a very high level due to the general increase in the world population and the per capita increase in water usage concomitant with a developing economy. Certain types of water use are consumptive, while water from other uses can be reutilized after purification. Table 4.1 describes the major uses of water.

As countries develop, not only does the total demand rise, but the relative importance of various types of use also changes. The supply to meet this

Table 4.1 Types of water use

Type of use	Characteristics
Municipal: potable water	Consumptive.
Municipal: other uses	Usually heavily polluted after use. Possibility of reuse after proper treatment.
Industrial	Various uses, a large proportion for cooling purposes. Level of pollution depends on industry involved. Substantial portion could be reused after treatment.
Hydro-electricity generation	Almost entirely reusable. Some evaporation loss from reservoirs.
Agriculture	Polluted by fertilizers, pesticides and saline concentration. Could be reused under favourable conditions after treatment. Otherwise usually ends up in local streams.
Navigation	All reusable. Level of pollution low.
Recreation	Reusable. Level of pollution low. Some pathogenic organisms might be present.
Propagation of fish and wildlife	Reusable. Level of pollution low.

demand could be from rivers, lakes or groundwater, depending on local conditions (case study E). For example, in relatively less humid areas, groundwater becomes an important source of supply. The growth of a large city often requires large-scale groundwater extraction or the piping of water from a source many kilometres away.

The attempt to meet the extra demands for water of a specific standard that arises out of progressive development could be difficult and costly, and may lead to large-scale endeavours which result in environmental degradation. For example, the huge dams and reservoirs that have been built in recent years are associated with certain ecological problems (case study F). There are several 300 km long reservoirs in Africa: Lake Volta behind the Akosombo Dam, Lake Kariba behind the Kariba Dam, and Lake Nasser behind the Aswan High Dam. Case study G indicates the level to which the water of one of the world's greatest rivers has been degraded by the sheer size of the demand for water for various purposes. The development of water resources carries with it the onus for careful planning and its implementation. Planning should go beyond the engineering feasibility study, and should examine the possible impact of the development on the environment, wildlife and society.

In the rest of the chapter we discuss some of the areas where problems might arise as a result of utilization of water resources in order to meet the increasing demand. The examples have been taken from three areas: the problem of water supply to settlements, the pollution of water due to population pressure, and the need for environmental planning for large-scale water utilization projects.

The supply of water to settlements

In the 1970s, the World Health Organization (WHO) carried out a survey on access to drinking water in the developing world. As expected, there was a great disparity in the availability of potable water between the urban and rural sectors. According to the survey, only 20 per cent of the rural community had access to potable water, a figure which rose to 75 per cent in the urban areas. These overall figures hid the huge disparities which were revealed when the data were broken down into smaller areas: figures ranged from 100 per cent in some urban settlements to less than 1 per cent in specific rural areas (Biswas, 1978). The total number of people without clean drinking water increased with time. By 1980, this number had reached about 2 billion; they were also without proper sanitation arrangements.

The UN General Assembly declared 1981–90 as the International Drinking Water Supply and Sanitation Decade to encourage national governments to provide clean drinking water and sanitation to all citizens. At the end of the decade, the number of the people without proper drinking water was estimated to be 1.2 billion and there was very little change in the number of people without proper sanitation (World Bank, 1992). Also, as before, tremendous disparities existed across countries. For example, almost all rural people in countries such as Malaysia or Mauritius have access to piped water, in contrast to the situation for many city dwellers in other countries such as Haiti or Congo. Large areas of the developing countries are still without safe drinking water or proper sanitation arrangements to prevent the pollution of water sources.

One of the better results of development has been the associated arrival of potable water, which reduces health hazards. The diseases associated with an unsatisfactory water supply are various types of gastro-enteritic diseases, such as cholera, typhoid, amoebic and bacillary dysentry, infectious hepatitis, and so on. Other diseases such as schistosomiasis, onchocerciasis (river blindness), trypanosomiasis (sleeping sickness) and guinea worm could be endemic where people have to go to a water course for their supply. Schistosomiasis, unfortunately, may increase with the development of water resources, primarily because of the spread of irrigation canals and ditches. This leads to the extension and year-round establishment of the habitat of snails which act as hosts to the schistosomiasis parasite. The spread of schistosomiasis with stable perennial irrigation systems has been noticed in Egypt, Sudan, Kenya and the Transvaal (South Africa) where the infection rate in the irrigated European farms was nearly 70 per cent above that of drier area (Biswas, 1978). It should, however, be mentioned that currently the WHO has predicted the eventual elimination of river blindness. However, unless precautions are taken, year-round water availability also leads to increased incidences of mosquito-borne diseases: malaria, filaria, yellow fever and arborivorous encephalitides.

Apart from the improvement in health, bringing drinking water to the user is extremely beneficial to the women of rural areas. It is not unknown for women in the Third World to walk several hours a day to fetch water. Bringing safe drinking water to the user frees them from this burden, and not only provides time for productive work, but also decreases the physical stress on the relatively less-nourished women members of the family, who tend to consume less at meals, leaving most of the food for the men and the children.

So far as the supply to urban areas is concerned, a city ideally should have a safe drinking water supply, enough water at adequate pressure for fire-fighting, and water to meet the industrial demands of the area. This requires a source, a treatment system and a distributory network. The requirements of a dependable water supply are twofold: ensuring the supply of water to meet the local demand, and maintaining the quality of water needed to meet the various types of specific demand within the total. The volume of water required by an urban settlement depends on the population, the ambient standard of living, the climate and the demand from industries.

However, many cities of the Third World have become extremely crowded because of development and the associated better job prospects than those in the rural areas. A number of city-dwellers are forced to live in unsatisfactory conditions with little or no access to properly treated water or sanitary facilities. This leads to contamination of water especially in the rainy season. In many cities, the poor without access to piped water or a standpipe are forced to buy water for drinking from street vendors at prices much higher than the municipal supply. The quality of such water is also at best dubious. The rapid spread of cities also results in widespread use of groundwater, which unless properly monitored may lead to environmental degradation. Shallow groundwater sources in urban areas with inadequate sanitary systems are often polluted. The growth of industries compounds the problem by contaminating water supplies with various synthetic chemicals released from the plants as industrial effluent.

Case study E

Supplying Jakarta with water

Jakarta, the capital city of Indonesia located on the north-western coast of Java, is a 500-year-old settlement, although part of its latter growth (as Batavia) is associated with the Dutch occupation of these islands. The early city grew up on a low coastal plain flanked by a set of alluvial fans on the landward side. Jakarta recently has seen a remarkable expansion. In 1945, when the independent Indonesia came into existence, Jakarta's population was about 600,000. Its current population of 11 million is expected to double by the year 2025, which will make Jakarta the fifth most populous city in the world. The municipal water supply system was originally planned for about half a million people.

Jakarta has now extended beyond the alluvial fans to the foothills on the landward side and across the coastal plains to the previously unoccupied swamps next to the sea. One of the problems of such growth is maintaining the supply of water to the residents of Jakarta (WRI, 1996). Municipal services are rather basic and the existing infrastructure falls

Case study E (*continued*)

short of expectations. In 1989, these were the sources of water supply to the residents of Jakarta:

- private standpipes and wells, used by about half of the city's population;
- water bought from private vendors: about a third;
- piped water: 14 per cent;
- rivers and canals flowing through the city and heavily polluted, used by a small fraction.

Two observations can be made immediately. First, the quality of water for most of the residents is not good. This particularly applies to the water bought from the vendors and that collected from rivers and canals which are heavily polluted by domestic waste water and industrial discharge. In spite of its lower quality, volume for volume the price of water bought from the vendors is much higher than the piped municipal water. It is the people of slums and squatter settlements, who do not have either water piped to their dwellings or access to a standpipe or well, who are forced to buy water from the vendors. The poor therefore not only pay more but also receive water of dubious quality. Even regarding the piped water, more than half of it is lost in transmission. The municipal piped water also has to be highly treated because of the low quality of surface water (WRI 1996).

Second, the growth of the city and the concomitant rapid increase in demand reduce both the quantity and the quality of the water available. The expansion of the city in the foothills interferes with the groundwater recharge zones and less water is available in the subsurface. Pumping out water in the coastal areas allows penetration by saline seawater below the drinkable groundwater. Saline contamination has been observed several kilometres from the coast and the water level in the coastal plain aquifer has dropped below sea level. Furthermore, as the groundwater is stored below Jakarta in sandy aquifers in old alluvial deposits, its withdrawal leads to subsidence of the ground surface which currently measures in several tens of centimetres. The groundwater is also locally polluted. The two major sources of pollution are domestic waste, which releases nitrates nd microorganisms, and industrial landfills, which release toxic material.

Implementation of large-scale projects

The implementation of large-scale projects designed to improve the availability of water resources is an integral part of the planned development of the developing countries. Multipurpose storage reservoirs constructed behind large dams impound water for irrigation, hydroelectricity generation, flood amelioration, municipal use and recreation. A network of canals will bring the water to demand areas. The projects, however, may give rise to certain types of environmental problems, especially if constructed without proper understanding of regional conditions. Proper planning requires long-term hydrological data, which are usually not available. Even short-term records of hydrological, geomorphic or biological characteristics of a river basin are often scarce. Though some of the gaps may be filled by synthetic hydrology or reconnaissance studies, the problem is not entirely solved, and the lack of information could have some unforeseen and unfortunate effects on the environment. The construction of dams, reservoirs and canals should be viewed not only as engineering problems executed on the basis of decisions taken on grounds of economic or political expediency, but also as large-scale modifications of the environment which may or may not have several undesirable side-effects.

If the catchment area is undergoing active degradation as a result of forest clearance or the spread of agriculture, or if the natural conditions indicate a rapid rate of erosion resulting from regional geology, high relief and intense rainstorms, the derived sediment would arrive at the reservoir at a rapid rate, leading to the reduction of its water-storage capacity. The upstream catchment area therefore requires careful land management. A considerable amount of sediment and water arrives in the river, not via the large tributaries, but through smaller streams and by surface wash. Land management over the entire watershed is therefore needed. If flood control is the prime objective, a number of small upstream reservoirs may be more effective than one large one downstream. If silting is a problem, the catchment should be left vegetated as far as possible.

The construction of a dam usually affects the channel downstream. The water released from the dam being low in sediment load has a tendency to erode and deepen the channel. The most commonly quoted example of this comes from the United States. After the completion of the Hoover Dam, the Colorado River degraded the channel below the dam extensively. The problem of the Colorado is a common one, and should be remembered at the planning stage because the lowering of the channel may result in difficulties with the intake of the river water for various purposes. A different problem affects the stream courses where the tributaries contribute a large amount of sediment load. Dam closure and the subsequent decrease in the volume of water in the downstream channel result in the raising of the river bed. Downstream channel alteration depending on the nature of the environment may result in difficulties with water intake, or in channel shifting, or in increased flooding.

Another possible side-effect that should be taken into consideration at the planning stage is the drowning of valuable agricultural or residential land, or the possible inundation of areas of great natural beauty. Reservoirs in the uninhabited tropical rainforest may also submerge the habitat of rare plants or animals. Dams may block the migratory paths of fish unless fish ladders are built at the dam site. Certain projects dramatically alter the environment, as was the case with the Aswan Dam (case study F).

The story of the Aswan High Dam and Lake Nasser illustrates many aspects of the implementation of a large-scale engineering project in the developing countries. The project is often implemented using imported technology and financial assistance. There is insufficient ecological information for a proper environmental assessment to be made and, unlike the Nile, hydrological data are usually not forthcoming. The project may become a political showpiece, and unless there is environmental degradation of high order, such problems may be little recognized.

It is not suggested that large-scale engineering projects for river basin development be abandoned. What is implied is the need for prior investigation, especially concerning both possible environmental impact and the reliability of the project site. This general rule is of increasing significance in the developing countries, as often the background information necessary for planning is very scanty, and political expediency requires a large engineering showpiece project. The lack of information and local experience can be disastrous. This is of special importance in the Third World as a large number of dams were constructed in these countries from the 1950s. Like the Aswan High Dam and the other examples listed earlier, some of these dams are huge. The pattern is continuing. In 1986, the Guri Dam was completed in Venezuela. It has the largest capacity for hydroelectricity production in the world but it will be surpassed when two dams are completed. These are the Itaipu Dam at the Brazil–Paraguay border and the Three Gorges Dam on the Yangtze River in China. Needless to say, most of these dams are controversial from the viewpoint of their impact on the environment.

Case study F

The Aswan Dam on the Nile

The 6,695 km Nile has the longest known history of water utilization (about 7,000 years), and as a result, the longest record of river flow in the world. Its water has been utilized for basin irrigation from the passage of annual floods for thousands of years, and from the early nineteenth century a number of barrages have been built across the river.

The Nile has two major sources (Figure F.1): the White Nile, whose headwaters rise in Equatorial Africa beyond Lake Victoria, and the Blue Nile, which has its source in Lake Tana in the Ethiopian highlands. The flow of the headwaters of the White Nile is well regulated by several lakes and the vast Sudd swamps of southern Sudan. The Blue Nile, however, is an extremely floodprone seasonal stream, and, along with its major tributary, the Atbara, brings down the sediment-laden annual flood of Egypt. The flood sediment, deposited over the floodplains and delta lands of Egypt, has made the lower Nile Valley an exceptionally fertile and populated area.

In the twentieth century several dams have been built across the White and the Blue Nile, each with certain beneficial contributions and some environmental problems. The biggest of the Nile dams is the Aswan High Dam, built in 1971 with Soviet assistance. It is a huge structure with a large reservoir behind it (Lake Nasser), the tail end of which stretches into Sudan. This showpiece engineering project has increased Egypt's potential arable land from 2.8 million ha to 3.6 million ha, and has been instrumental in creating and maintaining self-sufficiency in wheat and enabling the export of rice. The planned annual capacity of hydroelectricity is 10,000 million KW. Agricultural areas have been converted from basin irrigation dependent on seasonal flooding to perennial irrigation, and flood control measures have improved. Unfortunately, the project simultaneously caused some environmental degradation.

As might be expected from its location, Lake Nasser has a very high evaporation rate, which reduces its efficiency. The sedimentation rate is also high, and that not only reduces storage capacity but, more importantly, it prevents the fertile silt from reaching the agricultural areas

Case study F (*continued*)

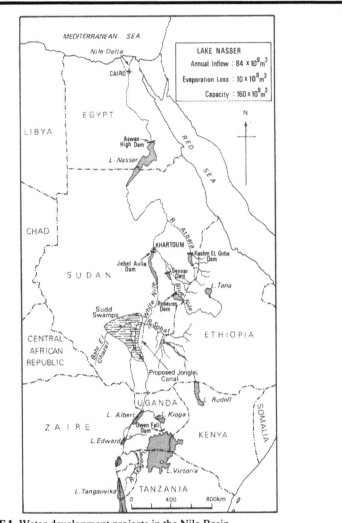

Figure F.1 Water development projects in the Nile Basin

Source: D. Hammerton, 'The Nile River: a case history', in R.T. Oglesby, C.A. Carlson and J.A. McCann (eds), *River Ecology and Man*, New York, Academic Press, 1972

Case study F (*continued*)

> downstream of the dam, forcing the cultivators to be dependent on fertilizers. The lack of water and sediment reaching the Mediterranean via the lower Nile has resulted in coastal erosion in the delta and also saltwater incursion. The lack of nutrients reaching the Mediterranean has destroyed the rich sardine offshore fishery which is partially compensated by fishing in Lake Nasser. The spread of schistosomiasis in the irrigated areas and salinization of part of the irrigated land have been even more disastrous (Hammerton, 1972).

Pollution of the waterways

All types of development – agricultural, industrial and settlement – create large amounts of detritus and polluted material which then find their way to the channels draining the area. Given sufficient time, rivers may cleanse themselves of some of the polluted material, but if the population pressure in the valley is high, such undesirable effects of development tend to persist. The river simultaneously becomes the way of pollution disposal and the source of drinking water. Case study G on the Ganga River is an excellent example.

The type of pollution found in rivers and lakes as a result of development, their sources, and the undesirable effects of such pollution are summarized in Table 4.2. Rivers have their way of dealing with organic wastes. When organic wastes from urban or industrial effluents or from agricultural fields enter the stream, the bacteria and protozoa which are present in the water utilize them as food in the presence of dissolved oxygen in the water. This demand for oxygen for degradation of organic wastes is referred to as biochemical oxygen demand (BOD). It is used as a pollution measure, and calculated as the amount in milligrams of molecular oxygen required to process 1 litre of polluted water in this fashion. As oxygen is removed from the water for processing organic wastes, its amount in the river water drops; this is known as the oxygen sag (Figure 4.2). Oxygen is then replenished from the atmosphere by a process known as reaeration, and, at the same time, as the processing of organic wastes goes on, the demand for dissolved oxygen drops. As a result, the oxygen in the water returns to the previous level, along with normal aquatic organisms. However, the replenishment of oxygen requires the water to travel over some

Table 4.2 Pollutants of water: types, sources and effects

Type	Major sources	Effects
Oxygen-demanding wastes	Sewage (animal and human), industrial wastes (food processing plants, paper mills, oil refineries, tanning mills), runoff from agricultural lands, decaying vegetation	Sag in dissolved oxygen, harmful effects on most aquatic organisms, foul odours
Agents of infectious diseases	Untreated domestic sewage and animal wastes	Spread of water-borne diseases, especially of gastro-enteritic tract, polio, infectious hepatitis
Plant nutrients	Domestic sewage, industrial wastes, phosphorus from detergents, runoff from fertilized fields, fossil fuel combustion	Algal bloom, excessive aquatic weeds, degraded taste and odour of water, oxygen depletion
Synthetic organic compounds	Fuels, plastic, fibres, detergent, paints, pesticides, herbicides, food additives, etc.	Possible toxic effect to aquatic life, birds, and humans; possibility of genetic defects and cancer; algal blooms and aquatic weed growth; foul odour
Inorganic chemicals and minerals	Mineral acids, inorganic salts, and finely divided metals or metal compounds from mining and industry; arsenic, petrol with lead, urban runoff	Increases acidity, salinity and toxicity of water often with disastrous effects
Oil	Machines, vehicles, tanker spills, pipeline breaks, etc.	Disruption of ecosystems and aesthetic damage to the environment
Sediments	Poor agricultural practice, destruction of forests, mining, construction activities	Filling of channels and reservoirs; increase in turbidity of water; reduction of certain members of aquatic population
Radioactive material	Natural sources, uranium mining, nuclear power generation and weapon testing	Cancer and genetic defects
Heat	Cooling water released from industries and power plants	Harmful effects on aquatic life, even fatal to some species; water solubility of oxygen decreased; increased chemical reactions

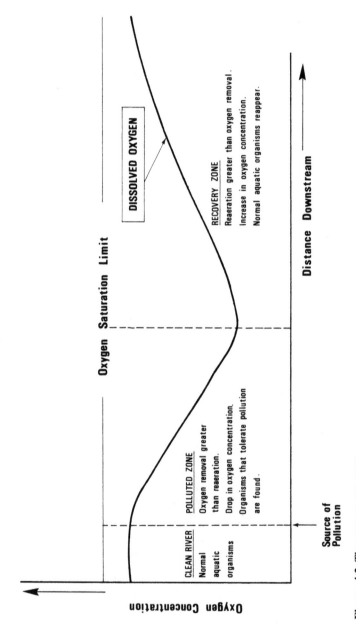

Figure 4.2 The oxygen sag curve

distance, and the addition of polluted matter within this distance would be undesirable.

In general, regarding the municipal and industrial sources of pollution, it is desirable to treat the effluent to an acceptable level before discharging it into the river where it can be further diluted. Unfortunately, this does not always happen. This problem is not restricted to the Third World, as many instances of irregular industrial discharge to the Rhine indicate. It is also possible to clean up a polluted river, as has been shown by the work on the Thames or the Singapore River, but treatment is expensive, and in the developing countries the cleaning up of a river is not necessarily perceived as progressive in the same way as the building of an industrial plant. The limitations of resources and awareness, the low priority given to environmental protection, and the power of the industrial establishments determine that the pollution of waterways will continue.

Case study G

The Ganga River in India: a case study of a river in use

The Ganga is 2,525 km long, and drains a 1 million km^2 basin (Figure G.1). Over 80 per cent of the basin is in India, where it constitutes more than a quarter of India's total land area and carries a quarter of its water resources. The river rises in the Himalayas from a glacier snout at 4,100 m elevation, but for most of its course, it flows through a vast alluvial plain at low elevations. The majority of the tributaries are both snowmelt and rainfed like the Ganga itself, although most of the south bank ones are entirely dependent on rainfall. Rainfall over the basin is controlled by the south-west monsoon, about 70 per cent of the annual rainfall arriving in four months (June–September). The river discharge thus shows considerable seasonal fluctuation.

Most of the basin, especially the part in the plains, is under agriculture. About 500,000 km^2 of the area is cultivated, some double or even triple cropped. Forests occupy only 14 per cent of the area, against the desired norm of 33 per cent of the Indian national forest policy. Forest depletion is common even in the steep mountains of the upper basin. Considerable erosion of the basins and a high sediment load in the stream are common, especially at the beginning of the rainy season.

Approximately 240 million people (more than the population of the

Case study G (*continued*)

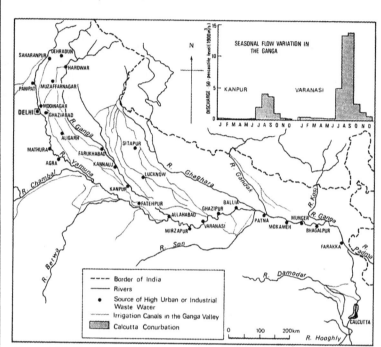

Figure G.1 The Ganga Valley

USA) live in the basin. The basin is highly rural, but still the urban population amounts to nearly 40 million people in more than 600 towns and cities. The distribution of urban centres is uneven, being higher in parts of the upper Ganga Valley and at the lower end in West Bengal. A large number of urban settlements, some of considerable size (examples: Ghaziabad, 226,000; Aligarh, 200,000; Kanpur, 1.5 million; Allahabad, 590,000; Varanasi, 719,000; Patna, 623,000; Calcutta conurbation, about 10 million), are located directly on the banks of the Ganga. The water of the Ganga is used mainly for:

Case study G (*continued*)

- *Irrigation* – a network of canals takes out large volumes of water to sustain agriculture in the valley. This leads to a drop in discharge for some distance below the canal headworks;
- *domestic use* – both urban and rural settlements use the river as a source of water for drinking and other domestic uses, including sewage disposal;
- *industrial use* – the river supplies water to the industries located mainly in the urban areas of the valley, and especially in the two concentrations of urban settlements mentioned above. The industries on the bank include tanneries, petrochemical and fertilizer complexes, pesticide factories, rubber, jute, textile and paper mills, and distilleries;
- *disposal of waste water* – waste water from agricultural fields, cities and industries is released into the river, and not always after proper treatment;
- navigation;
- *religious practices* – the water of the Ganga is holy to the Hindus and the river is extensively used for bathing. On certain holy days of the year, millions of pilgrims bathe at certain points along the river. The river is also used for the disposal of human bodies for the same reason.

The nature of the physical environment and the high and various demands on the water lower the ambient quality of the river water. Untreated urban waste water is the prime polluter of the Ganga, followed by industrial effluents. The untreated waste water arrives even from large cities like Varanasi, and such point sources of extreme pollution make the passage of the river through the states of Uttar Pradesh, Bihar and West Bengal hazardous. The organic pollution from rural areas is widely distributed, and probably does not constitute a hazard except in very densely settled areas. The agricultural waste water also causes pollution problems, especially from fields with high applications of fertilizers and pesticides.

The problem of maintaining the quality of water is accelerated during the low season, and also downstream of the points of water withdrawal for irrigation. The quality fluctuates depending on the density of population along the banks, the location of urban centres, and the location of tributary junctions. For example, in the reach between Kanauj and

Case study G (*continued*)

Kanpur, the BOD rises to the high level of 10–20 mgl⁻¹. The colliform bacteria count is also rather high.

A regular programme for monitoring water quality in the Ganga, which involved sampling at various stations along the channel at regular intervals, started in 1979. In order to improve the general quality of the river's water, it is necessary to control the land use in the basin, the discharge of wastewater effluents (both domestic and industrial types), and other intensive uses of the river. An integrated plan and considerable effort and resources are necessary to achieve this. A Pollution Control Research Institute was established at Ranipur, near Hardwar in the Himalayas. The Ganga clean-up project (the Ganga Action Plan) was officially launched by the then Prime Minister, Rajiv Gandhi, in June 1986 at Varanasi, and financial and technological assistance was available from overseas donors and the United Nations Development Programme. At the beginning there were some encouraging signs, such as the formation of the local action groups like the Swatchh Ganga Abhijan of Varanasi and the successful overhauling of the sewerage system of the holy town of Hardwar so that the bathing pilgrims were spared the upstream disposal of the town's waste water. Over time, however, the river has not shown much improvement, rather the opposite. The clean-up project has been accused of misusing funds, overspending, bad planning and slow progress. It is imperative that the urban sewage and the industrial toxic effluents which extensively pollute the Ganga, without giving it enough time to reaerate itself, are brought under strict control and proper waste treatment before release into the river. It is undoubtedly a massive enterprise and requires forcing the river bank industries to treat their waste water properly and the hundreds of cities in the basin to deal with their municipal waste water adequately.

Key ideas

1 Demand for water increases directly with rise in population and with economic development.
2 Certain types of water use are consumptive, while water from other uses can be reutilized after treatment.

3 Large areas of the developing world are still without safe drinking water or proper sanitary arrangements. One of the best results of development is the arrival of safe drinking water, which reduces health hazards.

4 Large-scale water development projects bring both benefits and environmental problems.

5 Background information and some local experience are necessary for the successful planning of large-scale water development projects.

6 Water is polluted from a variety of sources: human sewage, animal wastes, mining, industry, pesticides, fertilizers, detergents. Rivers are usually capable of purifying themselves of a limited amount of oxygen-demanding wastes.

5

Development and changing air quality

The constituents of air

By air we refer to a mixture of gases that envelops the surface of our planet. Remarkably the composition of this mixture is nearly constant from the ground level to a height of 80 km. The two major constituents of air are nitrogen (78 per cent by volume) and oxygen (approximately 21 per cent). Argon and carbon dioxide are also present but together they constitute only about 1 per cent. The remaining major constituent, water vapour, fluctuates between 0.01 and 5 per cent. A large number of gases, the minor constituents of air, make up about 0.01 per cent by volume. When some of these are present in sufficient quantity to affect the physical well-being of humans, animals, vegetation and materials, they are considered pollutants of air. Such pollutants may exist as solid particles, as liquid droplets, in a gaseous state or as a mixture. Usually if air is polluted, it is simultaneously by more than one kind of pollutant. Local concentrations of such pollutants, as are often found in industrial and urban areas, are accentuated by meteorological and topographic conditions (Figure 5.1) which prevent the mixing of air and the associated dilution effects. This occurs, for example, over Mexico City and Kuala Lumpur.

The spectre of air pollution hangs over industrial and urban centres. The air pollution over London in the 1950s and several disastrous instances of pollution over New York in the 1950s and 1960s are textbook examples. Both London and New York are now much cleaner, as far as their air is concerned, due to concerted efforts put in after such disastrous episodes. Unfortunately, as the developing countries go through industrial and urban development, such attempts to keep the air clean have not yet started to happen.

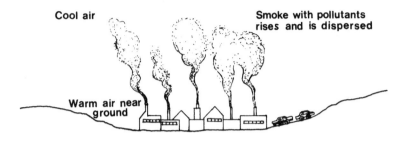

NORMAL CASE OF DROP IN TEMPERATURE UPWARDS

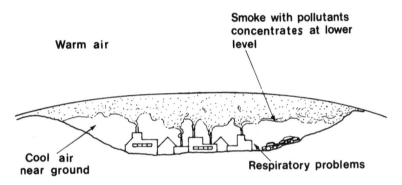

TEMPERATURE INVERSION IN A VALLEY

Figure 5.1 Inversion of temperature and air pollution

The major pollutants of air are oxides of sulphur and nitrogen, carbon monoxide, hydrocarbons, photochemical oxidants (ozone, organic aldehydes and peroxyacyl nitrates or PANs) and particulates. Particulates are also known as suspended particulate matter (SPM). The major sources of pollutants are various modes of transportation, stationary fuel combustion, industry and disposal of solid wastes (Miller, 1986). The concentration of the pollutants of air is measured either in parts per million by volume (ppm) or as the mass of pollutants in microgrammes to the volume of containing air (μgm^{-3}). The details of the agents, sources and effects of air pollution are listed in Table 5.1.

Development in rural areas and air pollution

Air pollution in rural areas is mainly from the burning of natural vegetation and wind action on bare fields. Large-scale clearing of the forests for

Table 5.1 Air pollution: types, sources, effects and controls

Type	Source	Effect	Controls
Sulphur oxides	Combustion of sulphur-bearing coal; petroleum products; industrial processes, particularly those involving smelting of sulphide ores of copper, zinc and lead	Acid rain damages buildings and vegetation; accelerated corrosion of materials; affects the respiratory system; high concentration (10 ppm) causes eye and throat irritation	Reduction in use of sulphur-bearing coal; clearing coal of some sulphur compounds; limestone injection in furnaces; use of alkalized alumina
Nitrogen oxides, mainly NO and NO_2	Combustion of fossil fuels; transportation	NO_2 concentration in air affects respiratory tract	
Particulates	Stationary combustion in households and power plants, especially those using coal; motor vehicle emissions; refuse incineration; burning of vegetation; industry such as blast furnaces and smelters	Respiratory diseases; possibility of lung and stomach cancer, especially if toxic material is present in particulate form; particulates between 0.1 to 5 μm most likely to affect humans; loss of visibility; increases the effect of other pollutants	Less use of coal for power generation; location of power plants away from densely settled areas; controlled incineration
Hydrocarbons and photochemical oxidants including ozone, aldehydes and PANs	Motor vehicle emissions; solid waste disposal by incineration; petroleum and chemical industries; rubber and plastic manufacturing; evaporation of organic solvents	Respiratory difficulties; eye irritation from aldehydes and PAN compounds; nose and throat irritation from ozone and aldehydes; photochemical smog; plant damage	Better engines and more economical use of fuel
Carbon monoxide	Incomplete combustion of carbon in fossil fuels, and car exhausts; forest fires; petroleum refining; industrial activities	Haemoglobin in blood picks up CO thereby reducing its oxygen-carrying capacity which affects the central nervous system and psychomotor; cardiac and pulmonary functions; people with emphysema and heart disease highly vulnerable	Control of automobile exhausts; more use of large public vehicles
Lead	Automobile emissions	Brain damage; behavioural changes	Conversion to unleaded petrol
Noise	Transportation vehicles; urban areas; industry and mining; construction activities	Hearing loss; psychological effects	Noise control by better equipment, by buffers such as rows of trees, or by legislation

agricultural expansion, especially by burning, may even result in temporary air pollution over nearby cities as sometimes happens in South-east Asia. In rural areas, if cooking is done over firewood in an enclosed space, it subjects the cook to an extremely high level of air pollution. This happens also in urban settlements of developing countries.

Mining, industry and air pollution

Particulates in concentrated form pervade the air in mining areas. The pollution is caused by blasting, which releases particulates and noxious fumes, and by wind blowing across open-cast mines, across waste heaps, and across open dumps of toxic mined products like asbestos. Extraction of mineral resources is an early and common stage in the development of the Third World countries, where safety rules and regulations are not always rigorously imposed.

Mine workers are often expected to work in appalling conditions for long periods at very low pay. Safety measures are uncommon, and the incidence of accidents is high. Respiratory diseases and eye ailments are frequent. The clustering of stone quarries and crushers leads to a thick pall of silica dust over the region, prolonged inhalation of which may cause silicosis with irritant cough, shortness of breath and chest pains. About ten years ago, Agarwal and Narain (1986) gave some examples from India of such pollution and its effect. They referred to a survey of an area near Delhi where 17 per cent of the workers were incurably affected. Dust from mining also impairs agriculture in the nearby fields, as seen around the magnesite ($MgCO_3$) mines of Jhiroli in Almora District of the state of Uttar Pradesh in India. One-third of the workers in the mica mines of Bihar suffer from silicosis. The worst case in India could be the slate pencil factories of Mandsaur in Madhya Pradesh. The pencils are made by using a cutter on a block of slate in a profusion of silica dust. A large portion of the work-force dies before the age of 40, and it is difficult to find a woman who has not been widowed at least once (Agarwal and Narain 1986).

Silicosis can be avoided by dust control and regular medical examinations, but such measures are not taken in poor and remote areas. This is not an isolated instance: situations of this type occur across the developing world, where poverty and desperation force the workers to take up the sole job opportunity in a remote and poor part of the country, impelling them to work towards an untimely death. The only way to control such exploitation is to raise the consciousness of the people at both the national and local levels about the hazards of air pollution. Asbestos mines provide even grimmer examples, where fine fibres of asbestos are deposited in the lungs, causing pulmonary fibrosis, which ultimately leads to respiratory difficulties and death. The alternative is possible cancer of the lungs or gastrointestinal tract.

As countries develop economically, their power requirements rise, and in places this need is met by burning coal to produce electricity. If the coal contains a high amount of impurities, the emission of the power plants becomes a continuous source of air pollution. The burning of sulphur-bearing coal, as for example in China, results in the emission of sulphur dioxide into the atmosphere, which if released in a considerable amount may cause acid rain. In countries with sulphur impurities in coal, acid rain has destroyed vegetation catastrophically and has damaged buildings and statues of historic interest. Even without sulphur impurities, the problem of air pollution from burning fossil fuels is accentuated when very large power stations are built, like the superthermal power stations of India with an ultimate capacity of more than 2,000 megawatts. The establishment of such super power stations is justified on grounds of economic location, but such economics does not necessarily take into consideration the fact that super power stations are also super sources of air and water pollution. The emission of fly ash with toxic minerals is also potentially a hazardous product of coal-based power stations. In many places, the thermal plants either do not have pollution control measures or the existing equipment does not function adequately.

Case study H

Effect of local air pollution on the Taj Mahal

The case of the Taj Mahal is an example of what could be inadvertently destroyed in the course of economic development unless a constant vigil is maintained. This famous mausoleum (Plate H.1) is made of Makrana marble of Precambrian age, and of course is susceptible to acid corrosion. Several years ago, a petroleum refinery was built at Mathura, 40 km away from the Taj at Agra. Several industrial ventures were also started at about the same time in Agra or in the neighbourhood. As a result air pollution increased and concern was expressed regarding the preservation of this beautiful building. The central government stepped in, a study committee was instituted, and certain of its recommendations were implemented. Two large thermal power stations and coal-burning locomotives have been removed from Agra. The sulphur dioxide over the city dropped by 75 per cent by the mid-1980s (Agarwal and Narain,

Case study H

Plate H.1 The Taj Mahal

1986). The emission of sulphur dioxide and the particulates from the refinery at Mathura is monitored regularly and, given the general pattern of wind movement, is said not to be a serious environmental threat to the monument. Some of the recommendations of the committee have not been fully implemented yet, such as moving 250 or so small iron foundries out of Agra, and planting a green belt around the city as a buffer against possible air pollution from distant sources. The attempt to declare a no-industry zone in Agra unfortunately has not been kindly received locally.

This example illustrates two characteristics of industrial pollution. First, it is all-pervasive, and second, even if technological solutions exist, their implementation is not easy. The damaged historical monuments of the developed world, where industrial growth and accelerated air pollution made an early appearance, clearly show the results of this.

With the development of power plants in the developing countries, a concentrated settlement of workers usually grows up in the immediate vicinity of the power station, close to the source of pollution. The combination of untreated emission of polluted matter from the industry and the concentration of labour in the neighbourhood is extremely common in the developing countries that are trying to build up an industrial base. Blast furnaces, for example, emit particulates and toxic fumes, but probably very few of them have any kind of pollution control measures which also function adequately. Other industries as listed in Table 5.1 tend to emit quantities of polluted matter into the air, and frequently are located uncomfortably close to an urban settlement, the growth of which may be connected with the industrial base.

Air pollution in cities

The cities of the developing countries have brought together industrial establishments, congested transport systems and concentrations of people. Such a juxtaposition results in various types of ailments of the lower and upper respiratory systems, chronic heart and lung diseases including bronchitis, fibrosis, cancer, and possible eye, nasal and skin irritation. The pollution of air in these cities is the combined effect of burning coal or firewood, motor vehicle emissions, power station combustion and the emissions from various types of industries. The developed world is converting to unleaded petrol, catalytic converters in cars which burn the carbon properly and avoid the emission of carbon monoxide, and, in the near future, the use of electric automobiles. Only a handful of developing countries have done or are able to do so.

The problem is becoming more acute in the developing countries because of the progressive building of industrial bases, the rapidly developing cities, and the in-migration of the poor to squatter settlements which because of the prevailing poverty are located within walking distance of the workplace. This is a high-risk situation, especially where people are uninformed, where the industry is callous about environmental degradation, and where the government is indifferent or inefficient. Hazardous forms of air pollution may happen anywhere, but in the Third World much of it happens due to neglect of safety measures. Highly toxic and volatile chemicals are stored dangerously; there is little monitoring of the health of workers; and pollution safety measures, such as scrubbers, do not always arrive with the new plant. Such practices are of course found elsewhere, but since industrialization is becoming a part of the economy of the developing countries, sensible precautions for the preservation of the environment should be built into the development plans, thereby avoiding both the industrial pollution and the subsequent massive clearing-up operations of the developed countries. The more heavily industrialized nations of the Third World are already showing signs of regional concentrations of polluted environment.

Although very little hard data exist for air pollution over cities of the developing countries, such air usually contains particulates well over the acceptable limit set by the World Health Organization (WHO). Other common pollutants that occur in high amounts are carbon monoxide and lead. Sulphur dioxide and low-level ozone have variable distributions. The figures are increasing because of three factors discussed earlier: rapid population growth; industrialization; increased energy use. A study carried out by the World Bank indicates that bringing down of the concentration of the particulates to the level acceptable to the WHO would annually avoid between 300,000 and 700,000 premature deaths in the developing countries. Given the fast rate of urbanization of the developing countries and the rapid growth of the very large cities, urban air pollution is likely to become an even bigger problem, especially for the urban poor.

Case study I

The disaster at Bhopal

The city of Bhopal, with a population of about 800,000 (Figure I.1), is the capital of the state of Madhya Pradesh in central India. In the 1970s, Union Carbide established a factory in the northern part of the city, which among other things produced chemicals for pesticides and stored them after production. It was a popular move as it meant jobs for people in town, and provided a means for meeting part of the rising demand for pesticides after the Green Revolution. In the early 1980s, there were several leakages and accidents in the plant: warning articles were written in the local press, and representations were made at various levels of the government, but apparently the warnings were not taken seriously by either the plant management or the local government.

About 11.30 p.m. on the night of 2 December 1984, workers in the Union Carbide plant noticed a gas leak. The leaking gas was methyl isocyanate (MIC), which is used in pesticide manufacturing. The gas leaked out in horrendous quantities. The safety measures were inadequate; and inexplicably, the public warning siren of the factory was not properly sounded. The highly poisonous MIC gas spread in high concentration over 40 km^2 affecting about 200,000 people. People woke up coughing, breathing became impossible, and they started to flee in whatever mode of transport was available: cycles, bullock carts, auto-

Case study I (continued)

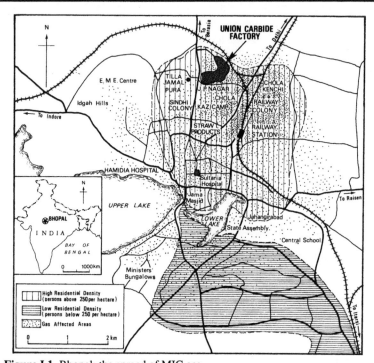

Figure I.1 Bhopal: the spread of MIC gas

Source: A. Agarwal and S. Narain (eds), *The State of India's Environment, 1984–85*, New Delhi, Centre for Sciences and Environment

rickshaws, cars, buses, trucks. Entire families in desperation travelled on a single scooter. People rode on the outside of jam-packed trucks hanging on to the limbs of people inside. The city streets were jammed with an unending moving stream of humanity. People who found a vehicle to escape on survived, but a very high percentage of those who fled on foot or stayed in their homes did not. The number of people who died is still not known exactly, but the total climbed into thousands. If it were not for the fact that the two lakes of Bhopal absorbed a large amount of the MIC gas, the number would have been even higher. Particularly badly affected were the poor who lived in shanties near the factories, in houses improvised out of planks, sheets of tin, plastic and thatch, with gaping

Case study I

holes that the gas could come through. Others were saved by the bravery of the various transport operators who evacuated thousands. The staff at the railway station close to the Union Carbide factory remained at their jobs, waved incoming trains through and alerted the neighbouring stations to stop trains from coming into Bhopal. Some of the railway staff succumbed to the gas themselves. Eventually units of the Indian army evacuated people systematically.

By the middle of the next day, 25,000 people were crowded into Bhopal's Hamidia Hospital suffering from eye and respiratory ailments. The problem was compounded by the lack of knowledge at that time regarding the specific gas responsible and the information was not forthcoming. The streets of the affected parts of Bhopal were full of vomit and human excreta; the dead lay in gutters; those who were alive suffered acute physiological distress. The city was full of dead animals and even the plants were not spared.

The long-term effect of this extreme form of air pollution is not known, but voluntary agencies working in Bhopal have reported that several thousand people are still suffering from respiratory, sleeping and digestive problems acute enough to make them incapable of carrying out even very light physical jobs. There is a possibility that these people, most of whom were poor manual labourers, will never be able to earn a living. Agarwal and Narain (1986) provide a harrowing account of the disaster in their edited book, *The State of India's Environment 1984–85*. More than ten years after the event the arrangements for compensation to the victims are still not satisfactory.

With hindsight, the Bhopal disaster could be attributed to several factors: the storing of a very large volume of an extremely poisonous substance; the safety system that did not work; the inexplicable failure to sound the public warning siren as soon as the gas leak was noticed; the late arrival of information to the doctors regarding the type of gas and its antidote; and allowing people to settle close to a hazardous plant.

If we extend the lessons of Bhopal to a general level, certain conclusions emerge. The needs of economic growth may require hazardous plants producing toxic fumes, but their establishment should be carefully planned and monitored. There should be definite policies for siting hazardous factories even though their presence may bring in jobs and economic opportunities in developing countries. Governments should not only have controlling legislation comparable to that anywhere in the world, but also display a willingness to implement it. The possibility of substituting less dangerous materials, a restriction on the amount of toxic material that may be stored, and the adequacy of safety systems in case of accidents should be carefully determined before the plant is installed. The developing countries need not carry toxic materials which are severely restricted in the developed countries. Factories which pose a danger to the community should have an uninhabited belt around them. It is not easy to implement such safety measures. The factories bring much-needed income to the local populace; and the poor in the developing countries tend to live as close as possible to the place of work in order to save transportation costs. An industrial plant run by a multinational has tremendous clout in a poor country. As always, coming up with technical solutions to an ecological problem is only a part of the total solution.

Air pollution standards and management

Air pollution standards are set to determine air quality. Crossing air pollution standards (acceptable limits) indicates that the air is legally polluted and it may have harmful effects. Standards can be used to evaluate the concentration of individual pollutants, or to measure the overall quality of the air, or to impose pollution limits for emissions released by industrial plants or thermal power stations. Air pollution standards are usually set by legislation in different countries and also by the World Health Organization. Major individual pollutants for which standards are set include carbon monoxide, lead, nitrogen dioxide, ozone, particulates and sulphur dioxide. However, the best known standard measures the overall quality of air over cities. It is known as the Pollutant Standards Index (PSI), and is often listed in newspapers for public information. The concentrations of the following pollutants are measured: ozone; carbon monoxide; suspended particulates; sulphur dioxide; particulates and sulphur dioxide in combination; nitrogen dioxide. The PSI is determined by identifying the worst of all these pollutants. Its concentration is then converted into a scale of quality with the following categories: good; moderate; unhealthful; very unhealthful; hazardous. This categorization is published in newspapers and public warnings are given when very unhealthful conditions are reached.

The major sources of pollution in urban areas are transportation, combustion in power plants, industry, and incineration including biomass burning. Three strategies exist for controlling pollution from these sources (Masters, 1991):

- *pre-combustion control*: usually accomplished by using cleaner fuels from which impurities have been removed;
- *combustion control*: improving the process so that less emission takes place;
- *post-combustion control*: pollutants are treated before their release.

For example, a thermal power station may use pre-combustion control by using low-sulphur coal or coal from which the mineral pyrite has been removed. Combustion control is practised when a mixture of powdered coal and limestone is used in the boiler which precipitates out the sulphur in coal as calcium sulphate. An example of post-combustion control is the use of scrubbers which spray a wet slurry containing finely powdered limestone into the gaseous emission. This also precipitates out the sulphur. The expected replacement of automobiles with internal combustion engines by electric vehicles in the twenty-first century should go a long way to make the urban air clean.

Key ideas

1 The major sources of air pollution are transportation, stationary fuel combustions, industry and disposal of solid wastes.
2 The major constituents of air pollution are oxides of sulphur and nitrogen, particulates, hydrocarbons and photochemical oxidants, and carbon monoxide.
3 Air pollution in rural areas is mainly from the burning of vegetation, wind action on bare fields and the burning of firewood, often in an enclosed space.
4 Mining is an important source of air pollution, as are thermal power stations.
5 Air pollution in many Third World cities results from motor vehicle emissions, power station combustion, the burning of coal or firewood, and industrial emissions.
6 The effect of air pollution is heightened by the tendency of the workers to live within walking distance of the factories, by the callousness of some industrial firms, by inappropriate legislation, and by the indifference of some governments.

6

Urban development and environmental change

Types of environmental modification

Cities are areas of the greatest alteration of the environment, areas where almost all the effects of ecological modification as a result of development come together. Cities have their own type of climate, vegetation and surface relief; their own brand of pollution; and their own specialized demands for physical resources. Our world is being progressively urbanized, the number of people living in the towns and cities is rising at a much faster rate than the world population, and increasingly an artificial environment is created, replacing the natural one. Urbanization can be defined as the proportion of the total population of a country or a region which lives in an urban settlement. The current rate of urbanization is especially high in the developing countries but is not so for the developed ones, although the percentage of people living in cities there is high from an earlier period of urban growth. By early next century, more than half the total population of the developing countries will be urban.

Although cities of all sizes are growing, the growth of megacities (those with a population exceeding 8 million) in the developing countries is extremely impressive. In the 1950s, most of the biggest cities were in the United States or Europe; currently such cities are located almost entirely in the developing countries. The projected future trends indicate that this pattern is going to continue. For example the fifteen largest cities of 1994 in order of their population were:

- Tokyo
- New York
- São Paulo
- Mexico City
- Shanghai
- Bombay
- Los Angeles
- Beijing
- Calcutta
- Seoul
- Jakarta
- Buenos Aires
- Osaka
- Tianjin
- Rio de Janeiro

Only three were in developed countries. The projected figures for AD 2015 indicate that only Tokyo and New York will remain in the list. The cities will also become huge. The projections show a Bombay of 27 million, a Lagos of 24 million and a Jakarta of 21 million. Metro Manila in fifteenth place is estimated to have nearly 15 million people (United Nations, 1995).

Such growth can only be accommodated by extension of city boundaries and alteration of the open space inside the current city limits, both leading to extensive environmental modification. The different types of environmental modification that result from urban development can be classified as:

1 climatological changes;
2 hydrological changes;
3 geomorphological changes;
4 vegetational changes;
5 an increase in different types of pollution.

In general, the quality of the environment deteriorates, and the various types of environmental modification occur simultaneously.

Climatological changes

The climate of a city differs in many aspects from that of the surrounding countryside. H.E. Landsberg (1981) has summarized our knowledge of this difference (Table 6.1). In general, the city, with its built-up areas and human

Table 6.1 Local climatic alterations produced by cities

Elements	Compared to rural environs
Contaminants:	
condensation nuclei	10–100 times more
particulates (dust)	10–50 times more
gaseous admixtures	5–25 times more
Radiation:	
total on horizontal surface	10–20 % less
ultraviolet, low sun	30% less
ultraviolet, high sun	5% less
sunshine duration	5–15% less
Cloudiness:	
clouds	5–10% more
fog, winter	100% more
fog, summer	20–30% more
Precipitation:	
amounts	5–10% more
days with < 5 mm	10% more
Temperature:	
annual mean	0.5–1.0° C more
winter minima (average)	1–2° C more
Surface relative humidity:	
annual mean	6% less
winter	2% less
summer	8% less
Wind speed:	
annual mean	20–30% less
extreme gusts	10–20% less
calms	5–20% more

Source: H.E. Landsberg (1981) 'City climate', in H.E. Landsberg (ed.) *General Climatology, 3*, World Survey of Climatology, vol. 3, Amsterdam, Elsevier

activities, produces enough heat by combustion, heating and metabolism to raise the average temperature by a few degrees centigrade. This pattern shows up clearly by drawing regional isotherms, and the term *urban heat island* has been coined to describe such a concentration of heat over urban areas. The heat island effect is more perceptible during daytime and in the first half of the night.

The widespread area, covered by buildings in close proximity increases the roughness of the city surface, which reduces the surface wind speed. The variation in roof heights results in turbulence. The wind circulation pattern of the city, however, is more difficult to establish. The urban heat island effect often results in a convective rise of air over the city, bringing in breezes from the countryside. This circulation pattern is strong at night, especially in calm, clear weather. The pattern of wind circulates dust particles, which are plentiful in cities, and a dust dome grows over the settlement resulting in the circulation of atmospheric pollutants over the city centre (Figure 6.1). If the city is located in an area where temperature inversion is common, such as a mountain-girt basin, or where fogs are prevalent, as off a cool coast, the pollutant concentration increases.

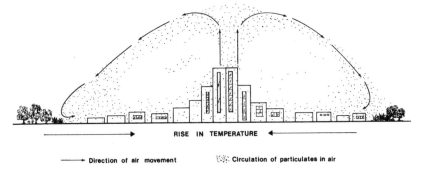

RISE IN TEMPERATURE

⟶ Direction of air movement   ⠿ Circulation of particulates in air

Figure 6.1 The urban heat island, air circulation and the dust dome

The atmospheric pollutants produced by a city are to a large extent solid particles suspended in air but also include gaseous constituents, some of which could have harmful effects. The solids are often hygroscopic, leading to the formation of fogs and reduced visibility, a characteristic which may persist due to the reduction of wind velocity inside cities. The concentration of some common pollutants of urban air is shown in Table 6.2.

Landsberg has described air pollution as one of the crucial environmental

Table 6.2 Concentrations of some air pollutants in city atmospheres

Pollutant	Concentration (ppm by volume)
Carbon dioxide	300 –1000
Carbon monoxide	1 – 200
Sulphur dioxide	0.01– 3
Oxides of nitrogen	0.01– 1
Aldehydes	0.01– 1
Oxidants, including ozone	0.00– 0.8
Chlorides	0.00– 0.3
Ammonia	0.00– 0.21

Source: H.E. Landsberg (1981) 'City climate', in H.E. Landsberg (ed.) *General Climatology, 3*, World Survey of Climatology, vol. 3, Amsterdam, Elsevier

Note: This happens with a 10–100 times increase in condensation nuclei, and 10–50 times increase in particulates

problems of today's urbanization, which may result in eye irritation, bronchitis, emphysema and asthma. Very little data exist for the cities of the developing countries, which in itself is worrisome, given the rate of urban spread and industrialization which often results in dense concentrations of population next to main city arteries and near industrial sectors.

Hydrological changes

Wolman (1967) has described three stages of hydrological modification of the environment, arising out of the spread of urbanization into the countryside:

Stage 1   The countryside before urbanization: land under natural vegetation or agriculture; streams adjusted to the existing conditions of the basin.

Stage 2   The brief period of construction: vegetation is removed; soil and weathered mantle undergo intense erosion; large amount of sediment is released to reach and partially fill drainage channels; streams in disequilibrium resulting from excessive sediment load.

Stage 3   The new urban landscape: construction completed, creating an impervious surface of streets, parking lots and roof tops; drainage via concrete drains; increase in flooding; streams still in disequilibrium, trying to adjust.

The degree of imperviousness in cities may range from nearly 20 per cent in low-density residential areas to about 90 per cent in the central business districts (Plate 6.1). The rainwater, instead of slowly infiltrating into the ground, rapidly runs off the rooftops, parking lots and streets into storm sewers and subsequently into major channels. Not only are infiltration and evaporation reduced in cities, but also as a result of quick accumulation of water, major channels, either natural or lined (Plate 6.2), rise more frequently in flood. The impact on the hydrology of the area from the urban spread can be summarized (Table 6.3).

Table 6.3 Modification of the local hydrology by urbanization

1 Increase in flooding: more frequent high flows in both concrete and natural channels.
2 Quicker rise in channels after rainfall.
3 Increase in channel peak velocity.
4 Decrease in the flow between storm runoff.
5 Generally a lowering of water quality and hydrologic amenities.

Plate 6.1 Singapore: intensive urban land use

Plate 6.2 An urban drainage channel

The general model of increased storm peaks and inter-storm low flows is accentuated in the tropics, as tropical rainfall often arrives in the form of brief, intense storms. In such a climatic environment, the flooding of the Bukit Timah Valley of Singapore has increased following the urbanization in the basin, where development tends to climb up the tributary valleys in the form of new housing. The urban flooding in certain Third World cities has worsened as a result of subsidence which has followed groundwater withdrawal, which was concomitant with city growth (Gupta, 1984). The best studied examples are Mexico City and Bangkok (case study J).

Case study J

Urban development, subsidence and flooding in Bangkok

Bangkok, which currently has a population of over 7 million, is located on the levee and the floodplain of the Chao Phraya River, about 25 km north of the Gulf of Thailand. Most of the metropolitan area extends across a low backswamp next to the levee of the Chao Phraya at an elevation of only 0.5 m to 1.5 m above mean sea level. Considerable rainfall arrives during the south-western monsoon, which is also the period when the Chao Phraya runs high. The city till now has depended on a series of canals constructed over the last 300 years for drainage into the Chao Phraya. Bangkok suffers from common developmental problems, such as pollution of water and air and expected flooding during the rainy season, but perhaps the greatest hazard at the moment is the effect of subsidence.

The city has grown rapidly since the 1950s, and the demand for water has been met to a large extent by tapping the groundwater resources. The water comes from several sand and silt aquifers within the soft marine sediments which include the highly compressible clay beds that underlie Bangkok. The water is for both domestic and industrial use, the industries finding it cheaper than the water supplied by the city. Such proliferation of unchecked pumping led to a rapid lowering of the water table, causing the shallow aquifers to turn saline to various degrees due to saltwater intrusion from the gulf. The real disaster for Bangkok was ground subsidence which has been carefully monitored, and found to be as much as 10 cm per year in parts of Bangkok (Figure J.1). The effect of this subsidence has been twofold: structural damage and increased flooding (Rau and Nutalaya, 1982).

Case study J (*continued*)

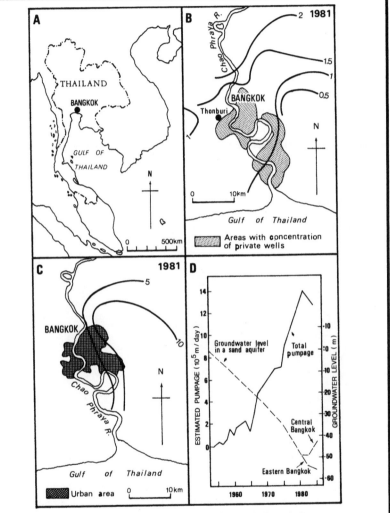

Figure J.1 Groundwater withdrawal and subsidence in Bangkok

Sources: J.L. Rau, 'Geotechnical problems in the development of Bangkok', *AGID News*, 1986, 49 (October): 30–4; P. Nutalaya and J.L. Rau, 'Bangkok: the sinking metropolis', *Episodes*, 1981, 3–7

Note: A – location; B – ground elevation; C – annual subsidence rate in cm; D – increased pumpage and falling groundwater level

Case study J (*continued*)

Structural damage commonly occurs at the junction of pavements with buildings. The large buildings are on piles and settle at a lower rate than the streets, resulting in the development of a prominent set of cracks at the contact. This leads to some of the characteristic features of Bangkok's urban landscape: large buildings require the addition of an extra step below the original set of steps leading to the entrance, sidewalks develop cracks and scarps, well casings protrude above ground surface, and walls open into large cracks one can put an arm through (Plate J.1). On the other hand, floods arriving regularly after rainstorms cause shop-owners to build ingenious small-scale embankments to prevent water from pouring into the shops from the streets (Plate J.2).

Plate J.1 Bangkok: structural damage from subsidence

Urban development and subsidence together have increased flooding in Bangkok. Several days of waterlogged streets are quite common there during the rainy season. In 1983, when the rainfall was exceptionally heavy, parts of the city were under water for months. It is no longer possible to drain off the water via the canals for various reasons, one of which is the subsidence-created depression of east Bangkok. A large number of pumps now move the water from one drainage channel to another, and

Case study J (*continued*)

Plate J.2 Bangkok: flood prevention measures

ultimately over the embankment into the Chao Phraya. The gradient of the canals has been lessened by subsidence, and they are also filled with city sewage. The floodwater that inundates Bangkok is extremely unpleasant.

Admittedly Bangkok's problems are somewhat extreme, but it is relevant to recall that a number of large and rapidly expanding cities of the developing countries are located on comparable substratum: Mexico City, Calcutta, Ho Chi Minh City and Jakarta are a few examples. If urban development requires supply from groundwater below the city, carefully planned and controlled extraction is necessary, as is now being attempted in Bangkok by stricter controls on groundwater pumping and by charging the consumer a fee which makes the costs comparable to that of the municipal supply. The water table has started to rise in central Bangkok but the problem remains very much in place.

Geomorphological changes

Urban expansion may give rise to two different types of geomorphological problems. The sediment associated with construction activities tends to block

RIVER BEHAVIOUR IN NON-URBAN AREAS:
Usually some water (baseflow)
between peaks associated with rainfall
events. Rise to peak after the rain has
fallen for some time, and a gentle fall
in river level afterwards.

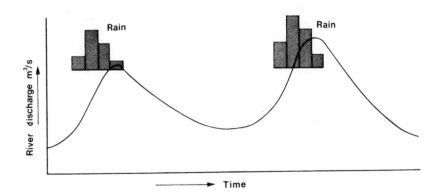

RIVER BEHAVIOUR IN URBAN AREAS:
Very little baseflow. Sharp rise to
peak soon after the rain has started,
and steep fall. Higher peaks.
Floodprone conditions.

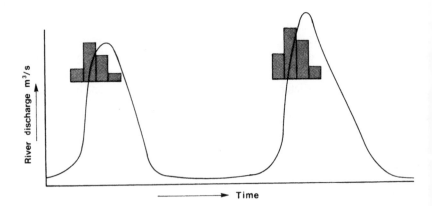

Figure 6.2 Comparison of hydrographs: rural and urban cases

the channels. This sediment finds its way to an urban river which is often an ephemeral stream with steep hydrographs, insignificant baseflow and large sediment concentration at certain times (Figure 6.2). During high flow the entire channel is under water, but between periods of high flow the stream is a braided one with shallow channels flowing in the middle of flood-deposited coarse sand.

Uncontrolled development up steep slopes may cause slope failures, especially when intense rain falls on a thick layer of weathered material. Landslides from intense rain have been reported repeatedly from Hong Kong, Singapore and Rio de Janeiro. Instances of fill slopes even being liquefied in heavy rain and failing have occurred. A standard scenario for such disasters is fairly common. Cities often develop on lower plains either next to or surrounded by steep hillslopes. Pressure on land forces settlement on such steep slopes. The poor may build squatter settlements and small farms on the steep hillsides, whereas the rich live in townships away from the polluted city air and with wide open vistas. Such steep slopes are often potentially hazardous due to a combination of unstable geological conditions and unfavourable geomorphic environment such as steep and active alluvial fans. The arrival of a tropical disturbance with intense rain acts as the trigger.

Cities built in naturally hazardous areas such as on the slopes of volcanoes or floodplains are, as expected, prone to disastrous landslides, floods, mud-flows, channel modification and destruction of people and property. The naturally hazardous areas include regions of high relief, seismically active belts, and areas exposed to the threat of tropical storms and cyclones. Some cities face all three as in Indonesia, the Philippines, Central America or the West Indies. The spread of urbanization in such areas has to be planned and monitored extra carefully.

Pollution in cities

The different types of pollution discussed earlier in this book – excessive sediment, dirty water, toxic air – are all found in cities, often in a concentrated form and in extremely close vicinity to the city-dwellers. The sediment is associated with construction activities, landslides or slope wash and usually affects specific areas within the urban settlements. Besides increased sedi-mentation and flooding, urbanization also affects the quality of the local water. The source of the pollution is the domestic and industrial wastes which reach the streams or groundwater. This happens across the world. Data from the developing world are rare, but given the large number of squatter settlements without a proper supply of drinking water or sanitary arrangements, the urban streams can be expected to be greatly polluted. A large number of urban settlements function without waste water treatment plants, and the polluted water is released directly into rivers or the sea. Where some data exist, even large rivers flowing through urban areas or the coastal waters in the case of seaside cities show a high level of pollution, a large proportion of which comes from untreated human and animal wastes.

The pollution of the air, as described in the previous chapter, becomes hazardous in the city because of the combination of high-density population and the concentration of pollutants. The concentration of airborne pollutants increases due to heavy traffic, industrial activity and combustion. The use of coal and firewood for cooking and heating in the Third World cities pollutes the air even indoors. Proper treatment of sewage and the supply of safe drinking water are extremely important goals for the developing countries. The twin problem of lack of safe and adequate drinking water and proper sanitary arrangements caused the United Nations General Assembly to declare the period 1981–90 as the International Drinking Water Supply and Sanitation Decade. As the cities in the developing countries grow and as industrial functions come in, it is imperative to monitor the water and the air of such cities, practices which are seldom carried out. The problems of the cities in the developing countries are rapidly increasing due to their brisk pace of urbanization. Because of the huge number of people involved, management of these cities may turn out to be as important a component of environmental management as saving the rainforest or learning to live on a warmer planet, as discussed in the next chapter.

Cities will certainly grow (both in population and area), but ideally such growth should be planned so that the negative impact of urbanization on environment is minimized. In certain instances urbanization may improve the quality of life. For example, urbanization could be associated with a regular supply of clean drinking water or access to medical facilities. If city planners are aware of the ambient physical environment at the outset, the deterioration of the physical environment as listed earlier in this chapter could be controlled. *Sustainable urbanization* is a current concept which calls for minimum deterioration in the physical and socio-economic environments of the city as it expands. Urbanization thereby is expected to take place without significant environmental degradation and the quality of life therefore could be sustained. This in practice requires considerable knowledge of the ambient conditions, good urban planning and efficient city management

Key ideas

1 The cities of the developing countries are fast expanding in area and size of population.
2 Cities are areas of the greatest alteration of the environment.
3 Urban rivers or drainage channels are floodprone.
4 Urban development on steep slopes, especially in areas with torrential rain or seismic disturbances, may cause disastrous landslides.
5 Pollution of land, water and air is common in urban areas.

7

The global concern

The spaceship Earth

As environmental degradation of the Earth grew, it became evident that a number of environmental problems were very large in scale. Certain types of degradation, such as large-scale air pollution, became a regional problem: in some cases, a continental problem. Transboundary air pollution from industrial sources and coal-burning power stations is an example. Some problems are even bigger, and global in scale. Such environmental misadventures are affecting or threatening to affect the entire planet. The two biggest are *stratospheric ozone depletion* and *global warming*. If they continue unabated, these two will affect every individual on Earth in one form or another. Their solutions require global efforts and co-operation.

The problems became global not only because of their scale but also due to the fact that the natural constituents of the Earth, such as water vapour or carbon, tend to circulate between their various storage places. The hydrologic cycle, discussed in Chapter 4, is an illustration. Human activities tend to alter the proportion of these constituents among the various storage places, leading to global environmental degradation. For example, carbon is stored in the rainforest. By destroying huge amounts of the rainforest and burning the biomass we have released quantities of carbon as CO_2 in the atmosphere. This in turn is threatening to change global climate. When that happens it will alter the quality of life for everybody irrespective of where the lost forests were located and for whom they were cut down. The solutions of global problems require stringent regulations and the need to involve everyone. The Earth behaves as a holistic unit in such cases.

Ozone depletion and its management

Ozone depletion

Ozone is a rather minor constituent of the atmosphere. About 3 units by volume of ozone are found in 10 million units by volume of air on average. Of this, approximately 90 per cent occurs in a belt at an altitude between 15 km and 25 km up in the atmosphere, known as the stratospheric ozone belt. This ozone concentration is also known as the ozone shield as it absorbs most of the ultraviolet (UV) part of the incoming solar radiation thereby making the surface of the Earth safer. Ultraviolet radiation tends to have harmful effects on people, plants and ecosystems. People exposed to UV radiation may develop skin cancer, cataracts and eye damage, and their immune system may be suppressed.

In the 1920s, a new industrial product, the chlorofluorocarbons (CFCs), started to be manufactured in the USA as a safe coolant gas in refrigerators. In composition CFCs include chlorine, fluorine and carbon. CFCs were neither inflammable nor toxic and chemically they were rather inert. Over time the use of CFCs grew enormously and different variations were put to different uses. CFCs have alphanumeric abbreviations depending on their chemical formula (Rosemarin, 1990). Apart from their use in refrigeration (CFC-11, CFC-12), CFCs were used also in air conditioners (CFC-12), open-celled plastics (CFC-11, CFC-12), car cushions (CFC-11), cleaning of silicon chips in electronics (CFC-113) and aerosol sprays (CFC-12). By the mid-1980s, the global production was over 1 million tonnes.

A very large percentage of the CFCs escapes from the containers and drifts upwards. Being chemically inert they move up through the atmosphere until they encounter the ozone shield. Above the ozone belt, the CFCs come into contact with solar UV radiation and break up to release the chlorine atoms. Atomic chlorine is highly reactive and almost immediately begins to combine with ozone molecules in a series of chemical reactions the end-products of which are an atom of chlorine and a molecule of oxygen. The chlorine atom reacts with another ozone molecule and the chain reaction continues. A single atom of chlorine can destroy up to 100,000 ozone molecules. Ultimately, chlorine is diffused down to the lower atmosphere and is returned to the surface of the Earth in rain. CFCs thus are extremely destructive to the ozone shield. Some of the compounds contain bromine. Such compounds (called halons and used in fire extinguishers) are even more destructive of ozone. Fortunately their use and production are limited. By their sheer volume CFC-11 and CFC-12 do most of the destruction of the ozone layer in the stratosphere. It takes decades for the CFC gases to reach the ozone layer (Rowland, 1990).

Ozone in the atmosphere can be measured in various ways: by ground-based instruments, from instruments carried aloft by balloons and airplanes and probably most effectively from satellites. Such measurements show a cumulative loss of ozone, especially over the temperate and polar regions. This has raised the spectre of strong doses of UV radiation reaching the Earth's surface with disastrous effects on people, plants and animals. The dramatic loss of ozone over the Antarctic has accentuated a feeling of impending crisis. Drastic drops in ozone over the other polar region, the Arctic, do happen but less consistently. Over the northern temperate zone ozone loss is high in winter. This being a well populated area, such a winter loss of ozone has given rise to considerable concern.

Restrictions on CFCs

By mid-1970s, CFCs had been identified as destroyers of the ozone shield by F.S. Rowland and M.J. Molina. It was therefore necessary to control the production of CFCs. Although they were mostly manufactured by the industrial nations, it was realized that the effect of ozone destruction was going to be worldwide. Again, as the developing countries grow more industrial they require more and more refrigeration, insulation and cleaning of electronic components. The arrangements for controlling CFCs therefore had to be global, both regarding production and usage substitution.

Some of the uses of CFCs are easily replaceable. For example, manual aerosol sprays replaced those using CFCs as propellants very early. Cardboard can be used for packaging instead of styrofoam. The other uses of CFCs, for example in refrigeration and air-conditioning, are harder to replace. Ideally, one should have a safe industrial product replacing the CFCs which would disintegrate in the lower atmosphere. At the same time the existing CFCs should be reused, and their escape into the atmosphere prevented as much as possible. Such substitution and alteration require funding, arrangements for which took about two decades starting from the original paper by Molina and Rowland (1974) which attributed ozone depletion to CFCs.

The late 1970s saw some restrictions imposed on the use of CFCs at the national level in a number of industrial countries. However, the required global restriction on CFCs finally emerged as a result of a series of international conferences, starting with Vienna in 1985, and also from progressively active encouragement provided by United Nations agencies and the World Bank.

The 1985 Vienna Conference for the Protection of Ozone Layer involved about twenty countries plus industrial and environmental organizations. The conference started international co-operation on controlling the CFCs. Between

Vienna and the next conference in Montreal two years later in 1987, a number of industrialized countries introduced piecemeal legislation for controlling CFCs at the national level. The globalization of CFC control, however, started with the acceptance of the Montreal Protocol on Substances that Deplete the Ozone Layer. This protocol required the production and consumption of five CFCs (11, 12, 113, 114, 115) and three halons to be frozen progressively between then and 1998. However, the Montreal Protocol was found to be inadequate regarding both restricting the amount of chlorine in the atmosphere and getting enough countries to accept the Protocol. These shortcomings were addressed at the next meeting in London in 1990, where about 60 countries, which between them produced 90 per cent of the world's CFCs, signed the Protocol.

A tighter restriction on the CFCs was proposed at the London meeting with acceptance of a plan for phasing out the five CFCs, halons and carbon tetrachloride. More concerted attempts were made to involve the developing countries in the Protocol by arranging a CFC fund for providing financial support to the countries of the Third World. The reluctance of these countries to agree to a ban on CFCs arose out of two considerations. First, the destruction of the ozone shield was essentially caused by the industrial nations who had produced almost all the CFCs currently in the atmosphere, and the developing countries felt that the industrial countries should be responsible for protecting the ozone shield. Second, replacing CFCs with new technologies could be expensive for the developing countries. A new fund, called the Global Environment Facility (GEF) was created to cover such expenses. Details of this fund are discussed in Chapter 8. Currently, the number of signatories to the Montreal Protocol is around 150.

The deadlines for phasing out the CFCs and associated chemicals were advanced at the Copenhagen Conference of 1992. It was agreed that the developed countries would phase out the CFCs by the end of the century and the developing countries by 2010. After some initial reluctance in the 1980s, the chemical industry including Du Pont (the world's largest manufacturer of CFCs) decided to produce CFC substitutes on a large scale. Such substitutes contain hydrogen or fluorine but not chlorine, and tend to break down in the lower atmosphere well below the ozone layer. Various industrial establishments in many countries currently manufacture home refrigerators and automobiles which do not use CFCs. Such appliances are clearly marked as non-CFC machines.

Currently, as a result of all these endeavours, CFC production is being phased out ahead of schedule. For example, the United States (the country that produced most CFCs) banned their production from 1996. However, the existing CFCs are still escaping into the atmosphere and the portion which is already there is making its way up to the ozone layer. The destruction of the ozone shield therefore will continue for several decades but there are signs that a gradual decrease of ozone-depleting substances is taking place in the atmosphere.

Case study K

Ozone depletion over Antarctica

The extremely rapid rate of ozone depletion over Antarctica gave rise to serious apprehension about the efficient functioning of the ozone shield. The apprehension arose primarily from measurements taken in 1985 by a British Antarctic Survey from the coastal station at Halley Bay. Such measurements taken with ground-based instruments showed a rapid decrease in ozone concentration overhead: from 320 Dobson units (DU) in 1960 to 180 DU in 1984. One Dobson unit is roughly equivalent to one part per billion by volume. In the next year US National Aeronautics and Space Administration (NASA) using instruments on the Nimbus-7 satellite found similar figures. This led to further investigation of the ozone layer by British, US and Japanese scientists using instruments based on satellites, balloons and aircraft. The general conclusions were that ozone had disappeared to a fraction of its late-1970 figures between the 15 and 20 km levels. The loss took place over an area as large as western Europe. The boundaries of the area with low ozone were clearly marked, showing a rapid loss within a very short distance. As a result this area of low ozone concentration came to be known as the ozone hole.

The loss of ozone takes place rapidly at the beginning of the Antarctic spring in September. Through the cold months of the Antarctic winter the air over Antarctica has a tight circulation, resulting in a large mass of very cold air in the stratosphere remaining in isolation. When the temperature of this air falls below −78°C, clouds made of crystals of nitric acid trihydrates or water ice begin to appear in the stratosphere. These clouds are called polar stratospheric clouds (PSCs). Chlorine occurs in the gas phase on the cloud surface. Everything waits in darkness through the cold Antarctic winter.

Spring comes in Antarctica in late August with sunlight. In the presence of sunlight the chlorine changes into atomic chlorine and immediately begins to react with the stratospheric ozone. The details of

Case study K (*continued*)

the chemical reactions over Antarctica are different from those which take place elsewhere, but the concentration of chlorine on the PSC surfaces leads to extremely rapid ozone loss and the hole appears every year with a lesser and lesser ozone concentration. This rapid destruction of stratospheric ozone stops by mid-October when the sun warms up the atmosphere enough for the PSCs to disappear, but the hole remains.

In 1987, this low ozone air drifted northwards to Melbourne, which caused considerable consternation. It is obvious that, if possible, the Antarctic ozone hole should not be allowed to expand. Similar ozone destruction happens over the Arctic, but only in certain years when the winter has been very cold. The discovery of the Antarctic ozone hole and the loss of ozone over the populated temperate latitudes in the late 1980s prompted the global conferences to phase out CFCs.

Global warming and its effects

Global warming has always occurred as a natural phenomenon but currently the state of our environmental degradation has converted it from a global benefactor to a global hazard. Global warming happens when part of the long-wave radiation escaping from the Earth through the atmosphere is trapped by a number of gases and re-radiated back to Earth thereby raising its temperature. The most effective of these gases is water vapour. The presence of water vapour and its global warming properties maintain the surface of our planet at an average temperature of 15°C instead of −18°C which would have happened if, like Mars, Earth had very little moisture in its atmosphere. Other global warming gases are carbon dioxide, ozone, methane, nitrous oxide and CFCs. The warming phenomenon is also known as greenhouse warming because of a rather tenuous analogy with the enhanced warming inside a glass-enclosed structure in the cooler climates, used for growing plants which need a lot of warmth. Global warming gases therefore are also known as greenhouse gases.

Global warming became a problem when the amount of greenhouse gases in the atmosphere, especially carbon dioxide, was increased by anthropogenic activities such as burning biomass, converting coal into electricity at power stations, and releasing CFCs into the atmosphere. The additional amount of these gases then produced excessive warming by more efficiently trapping the terrestrial radiation and sending it back.

Understanding the phenomenon of global warming involves the following:

- measurements of global warming;
- causes of global warming;
- effects of global warming.

Measurements of global warming

Even a few years ago the acceptance of global warming was not as widespread as it is today. Global warming is difficult to prove as temperature records do not go back very far. Furthermore, the old records are primarily land based, are not representative of large areas of the world, are mostly from urban areas, and are not always collected with precision. Existing records, however, were collated, processed and standardized by P.D. Jones and T.M.L. Wrigley (1990). Their standardized data indicate a slow warming trend since the last century with occasional periods of cooling. The deviations from the general trend may occur due to three reasons: sunspot cycles; volcanic eruptions producing large quantities of fine ash in the air; the occurrence of El Niño Southern Oscillation. Correcting for all such factors, Jones and Wrigley estimated that the earth has become 0.5°K warmer since the 1880s. Estimates by other researchers are comparable.

Evidences of global warming also come from other sources. In recent years, glaciers on mountains, particularly tropical mountains, have melted faster than before. The temperature of the top hundred metres of sea water off the coast of California shows an increase of 0.8°K over the last forty years. The data from the ice cores of Antarctica also indicate a warming trend. These cores through the ice indicate snowfalls of a number of years in sequence which later has turned into ice. As this happens tiny air bubbles get trapped in the ice, and these bubbles can be investigated to determine the composition of the air at the time of the snowfall and also the temperature. The latter is determined by examining the ratio of the two oxygen isotopes, ^{16}O and ^{18}O. The ratios reflect the ambient global temperature. It is also possible to date the sample.

A number of very hot years, in fact eight of the hottest on record, happened between 1980 and 1992. Apart from indicating the trend, this put global warming in people's minds. Currently global warming has become even politically an accepted ongoing event.

Causes of global warming

Svante Arrhenius of Sweden in 1895 demonstrated the linkage between carbon dioxide in the atmosphere and temperature. Carbon dioxide in the atmosphere has been recorded since 1957–8 at the Mauna Loa Climate Observatory in

Hawaii, thanks to the foresight of Roger Revelle who was the motive force behind starting the measurements during the International Geophysical Year of 1957–8. Data for earlier periods are usually collected from the air bubbles in the Antarctic ice. The recorded measurement of other greenhouse gases only goes back several years.

The concentration of carbon dioxide has increased steeply from the 1950s (Figure 7.1). Of all the greenhouse gases, it makes the biggest contribution (about 50 times more than any other gas) towards global warming. Water vapour is also an extremely efficient greenhouse gas, but its amount in the atmosphere does not increase. Gases such as methane or CFCs tend to absorb long-wave radiations at specific wavelengths, and supplement the warming mostly carried out by carbon dioxide (IPCC, 1994). If we wish to control global warming, lowering the emission of carbon dioxide to the atmosphere should be our prime target, although the emission of other greenhouse gases also needs to be controlled. Table 7.1 summarizes the background information regarding the greenhouse gases.

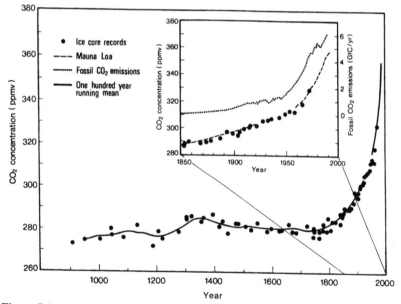

Figure 7.1 Rise in carbon dioxide concentration in the atmosphere over the last 1,000 years

Source: Generalized from UN Intergovernmental Panel on Climate Change (IPCC), *Radiative Forcing of Climate Change: The 1994 Report of the Scientific Assessment Working Group of IPCC. Summary for Policymakers*, IPCC, World Metereological Organization and United Nations Environment Programme (UNEP), 1994

Table 7.1 Greenhouse gases in the atmosphere

Gas	Atmospheric concentration	Anthropogenic source
Carbon dioxide	355 ppm	Biomass burning; energy production; agriculture; deforestation; industry, especially cement
Methane	1.71 ppm	Wet paddy fields; ruminating animals
Nitrous oxide	311 ppb	Agriculture; biomass burning; industry
Ozone, tropospheric	35 ppb	Chemical reactions in the atmosphere
CFC-12	0.50 ppb	Refrigerant; foam blowing; air conditioning
CFC-11	0.28 ppb	Foam blowing; refrigerant; solvents

Source: IPCC, *Radiative Forcing of Climate Change*, WMO and UNEP, 1994

Note: All concentration data are for 1992. Only anthropogenic sources are listed, not natural ones

Effects of global warming

A general agreement has emerged regarding the occurrence, rate and causes of global warming (Barron, 1995). The effects of global warming, however, are associated with uncertainty and controversy. Basically three types of approach are used to determine such effects:

- several computer-based numerical models are used to reproduce future conditions in a warmer Earth. Such models are known as General Circulation Models. These work at regional scales and for projecting general conditions. The models tend to agree on effects of specific climatic changes but not always;
- one can look for analogous situations in the recent geological past when the Earth was warmer;
- similarly, it is possible to use geographical analogues, that is, transposing conditions from areas which now have climate similar to the type being predicted for the area being studied.

Certain conclusions seem to have emerged from a number of studies. The current rate of average global warming is estimated to be around 0.5°K with the possibility of a small deviation. This rate is expected to rise by the middle of the next century due to the increasing presence of greenhouse gases in the

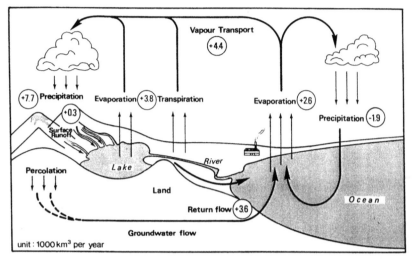

Figure 7.2 Projected effect of global warming from a doubling of the greenhouse gases for the period 2030–40

Source: L. Bengtsson, 'A numerical simulation of anthropogenic climate change', *Ambio* 1997, 26: 58–65

atmosphere. The increase in temperature, however, will vary both seasonally and latitudinally. Winter temperature in polar areas and both winter and summer ones in temperate latitudes will show the biggest relative change. Increased temperature may give rise to reduction of snow cover and sea ice, higher evaporation, greater precipitation (Figure 7.2) and increased storminess in the tropics. Details, however, are still being worked out.

A rising sea level could be the most striking of the consequences of global warming. At the beginning, the rise will be from thermal expansion of the sea water. A limited volume of water will be added by the melting of the glaciers on the mountains. The difficulty in predicting the new sea level lies in the behaviour of the ice sheets on Greenland and Antarctica. If the ice melts at a fast rate, as has been predicted for Greenland, the sea level rise would be much faster and higher. There is more uncertainty about the melting rate and volume of the Antarctic ice. The western part of the Antarctic ice sheet rests on a number of islands and frozen seas, not on land. A collapse and rapid disintegration of this part of the ice sheet is possible when the Earth is warmer. If that happens, the projected rise in sea level has to be modified upwards disastrously.

Currently, the mainstream projections of sea level rise by the year 2050 range between 5 cm and 40 cm. A rapid rate of melting of the Greenland ice and the collapse of the west Antarctic ice sheet will move these projections higher. As much as 3 m has been proposed as the disaster scenario but it seems likely that the world would not have to cope with a sea level rise of more than 50 cm in another fifty years. By then, better background information and estimates should be available.

Even a small rise will be problematical for the low-lying areas such as deltas and coral islands. Almost all coastal deltas will be submerged to some extent, and in the case of low flat deltas such as the Ganga–Brahmaputra system or that of the Mississippi, the area inundated will be a very large one. Low coastal plains, as in the Netherlands, Guyana or northern Australia, are also at risk. The problem is compounded by the fact that deltas and coastal plains are often densely populated, agriculturally rich and have urban settlements. Increased storminess in the tropics will increase the coastal hazard of submergence and erosion.

A number of coral islands such as Kiribati, Vanuatu and those in the Maldives have a very large proportion of their area under 1 m in elevation. A rising sea level, even of 50 cm, will be particularly disastrous for them. The reduction in size of the islands will make many uninhabitable. To this will be added loss of valuable coastal land for agriculture and coconut growing and for building tourist infrastructure. Increased salinity of the groundwater on which quite a few of these islands depend will worsen the situation. In the worst cases, the islands will have to be abandoned; and the scenario includes mass emigration of the islanders to safer places as *environmental refugees*.

Other possible effects of global warming are less clearly known and still require more research. It is, however, expected that the major rivers of the world will undergo adjustments to cope with changes in climate and sea level. Some rivers may become more seasonal and floodprone. Coastal erosion will increase and so will slope failures. The glaciers in the mountains will retreat rapidly. These physical changes will also have follow-up effects, such as shortage of water and electricity and changes in ecosystems. Vegetation zones may shift polewards when the climate is warmer or upwards into the mountains. If the rainfall pattern shifts it would also be followed by vegetational changes and faunal movement. Distribution of mangroves and coral reefs in the tropics and ecosystems at the desert edges could be particularly affected.

Other changes will involve soil moisture, temperature range and pattern of agriculture. Major changes in agricultural production would lead to national economic hardships or prosperity which in turn may have political implications. That is the most uncertain of all projected effects of global warming.

Case study L

Global warming and the Maldives

Six hundred and seventy kilometres to the south-west of Sri Lanka, the Maldives form a north–south chain of about 1,200 small coral islands atop a submerged mountain range in the Indian Ocean (Figure L.1). An atoll is a ring-shaped island enclosing a deep lagoon. In fact the word atoll came from the Maldives. The coral atolls in the Maldives are extremely low, less than 3 m in elevation, and range in size from very small coralline sandbanks to atolls whose longest diameter could be about 40 km. The land area of the biggest atoll is less than 13 km^2. Although the Maldives are affected by the monsoon storms, the outlying reefs protect the islands from intensive coastal erosion.

About 225,000 people live on a number of these islands with about a fifth living in Malé, the capital. The economy is dependent on shipping, fishing, gathering coconuts, and growing crops like millet and yam on tiny plots. Industry is limited to canning tuna and coconut products, handicrafts and boat building. Over the last twenty-five years or so, tourism has expanded rapidly in the Maldives, and currently is an extremely important part of the economy, utilizing the low atolls, the seascape and the beautiful coral reefs. The Maldives have become one of the famous tourist locations on the international scene.

Given the low elevation of the islands, even a moderate rise of 50 cm in the sea level will result in a tremendous loss of land for the Maldives. The usual litany of island problems are expected to happen such as coastal erosion, loss of agricultural land and tourist infrastructure, and an intrusion of saline seawater into the groundwater system, but it is the absolute loss of land which would make most of the islands un-inhabitable. Under these circumstances, a migration of population to higher islands or neighbouring land-based countries is to be expected; and with an even higher rise in sea level, the existence of the sovereign state of the Maldives could be threatened. In fact, almost the entire island chain could be under water.

The government of the Maldives has been deeply concerned with this predicament, and has initiated technical conferences on the consequences of global warming for the small islands. The President of the Maldives raised the issue at the 1987 Commonwealth Conference and also at the 1992 South Asian Association for Regional Co-operation (SAARC) meeting.

Case study L (*continued*)

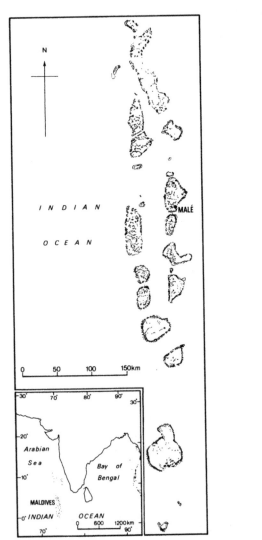

Figure L.1 The atolls of the Maldives: due to the small size and very low elevation of the scattered islands, most of the island group will be submerged if the sea level rises

Case study L (*continued*)

The case of the Maldives is an extreme one, but similar possibilities exist for a number of coral island states in the Indian and Pacific Oceans. There is, however, a school of thought which believes that the rising sea level would revive the coral growth, forming a barrier against erosion and submergence. Even additional sediment may be available to enlarge the existing islands. On the other hand, the only way to save these islands could be to construct embankments all round them. This would be very expensive, may not work and could lessen the tourist attraction considerably. An Intergovernmental Panel on Climate Change (IPCC) estimate suggests that protecting the Maldives would cost at least one-third of the country's GNP every year.

The global concern

Both stratospheric ozone depletion and global warming are potential disasters on a global scale. Stratospheric ozone depletion developed from the widespread use of an industrial product which was originally designed for safety as well as for other objectives. The hazard was identified and proven by careful scientific work, and after several years of controversy the danger of the destruction of the ozone shield by CFCs was widely accepted even by the producers of the chemicals. The next stage involved working out a global agreement for phasing out the CFCs plus producing and using substitutes. This again took about a decade.

Scientific research on global warming is still continuing, although a general consensus has emerged on its causes and major effects. The details, such as rates of global warming and sea level rise, are still being worked out; and, like ozone depletion, a global agreement on restricting the greenhouse gases was proposed and agreed by most of the countries. This happened at the 1992 environmental conference at Rio de Janeiro.

Both instances illustrate that seriousenvironmental dgradation may take place affecting the entire planet and we all may sufer from it, with or without making much contribution to the causes of the disaster. This necessarily requires global co-operation in many areas. Signing an agreement is only a beginning. It is essential to have good science and technology, political will and the willingness to provide financial assistance to the countries which are not in a position to acquire industrial substitutes. In both cases discussed in this chapter controversies arose between the developed and developing nations regarding responsibility and funding. It is obvious that global arrangements and co-operation are necessary to protect the environment of the planet.

Key ideas

1 The two biggest environmental misadventures are stratospheric ozone depletion and global warming.
2 Chlorofluorocarbons (CFCs) are responsible for progressive ozone depletion in the stratosphere.
3 Stratospheric ozone depletion leads to skin cancer, cataracts, eye damage and suppression of the immune system.
4 The worst case of stratospheric ozone depletion occurs over Antarctica.
5 International agreements now exist to phase out CFCs in the immediate future and provide substitutes.
6 Global warming is occurring due to the increase of greenhouse gases, particularly carbon dioxide, in the atmosphere.
7 The effects of global warming are still being worked out but they include climatic change, a rise in the sea level and associated problems.

8

Concepts and mechanisms for environmental management

Introduction

Rachel Carson's book *Silent Spring*, first published in 1962, probably contributed more than any other publication to initiating the current concern for environment. The book primarily deals with pesticide-related pollution. In the 1960s and even into the 1970s, environmental concern was primarily pollution-related. The pollution was perceived as deriving from anthropogenic sources such as thermal power plants, chemical industry, untreated waste water, marine oil spills. There were serious attempts at environmental management at the national level, for example, the passing of the National Environmental Policy Act (NEPA) in 1969 in the United States requiring environmental evaluation of new projects and the establishment of the Environmental Protection Agency.

Environmental concern grew into a popular movement which observed the first Earthday in 1970. A number of organizations drew attention to the progressively degrading environmental conditions. These organizations, being popular movements, were known as the non-governmental organizations (NGOs). The number of books on environmental issues proliferated, spreading awareness, although some of these now read very strangely. The Stockholm Conference in 1972 extended the national and local awareness to the global level, and related environmental management with economic and social development.

The Stockholm Conference, 5–16 June 1972

The Stockholm meeting was called by the United Nations with Maurice Strong as the conference secretary-general. Originally, it was planned to deal with pollution which was becoming transboundary in nature. It was felt that a global meeting was necessary to resolve the issue. However, the developing countries stressed the fact that global environmental management is only possible when environmental management is linked with development and the transfer of knowledge and technology. The conference globalized the environmental movement and the movement shifted focus to link development and environment together.

The main conference was preceded by a set of meetings which were necessary to resolve the differences among the countries prior to the meeting at Stockholm. The Stockholm Conference was attended by representatives from 113 countries, a number of international bodies and the NGOs. This was the first time NGOs had been invited as active participants in environmental management. However, Indira Gandhi was the sole political leader of importance to attend Stockholm. One oft-quoted remark of Indira Gandhi at the conference was that poverty is the greatest polluter. It was becoming clear that the protection of the environment is not possible so long as countries remain poor. It is not possible to tell a villager with no other access to energy not to collect firewood. Development requires the improvement of both economic and environmental conditions.

Stockholm was the first conference which effectively brought the global environmental community together. Although not all the expectations were fulfilled, the conference resulted in achieving the following:

- understanding of the *global* nature of the environment;
- recognition of the *linkage* between development and environment;
- strengthening the *environmental institutions and priorities* within the United Nations;
- accepting the role of *NGOs* in environmental issues;
- resolving important environmental issues in *conferences.*

All this was achieved without major confrontation, and the developments at Stockholm led to new institutions, policies and mindsets which have governed envirnmental management at various levels since then. In tht sense modern globl environmental anagement started in Stockholm. Quite a few of the organizations and programmes listed below came into existence because of Stockholm.

United Nations Environmental Programme

The United Nations Environmental Programme (UNEP) started in 1972 following Stockholm. The UNEP collects, processes and distributes information on the environment. For example, it carries the International Referral System (INFOTERRA) which is involved in the exchange of environmental information, and the Global Environmental Monitoring System (GEMS) which compiles environmental information at global and national levels after acquiring such data from various national governments. The UNEP is based in Nairobi, the first major UN agency to be located outside the United States or Europe.

The UNEP also acts in a co-ordinating capacity with a number of environment-related international organizations. The Convention on International Trade in Endangered Species of Wild Fauna and Flora (CITES) is an important example. The UNEP has been involved in administering the secretariats for the 1985 Vienna Convention and the 1987 Montreal Protocol, both aimed at reducing stratospheric ozone depletion. One of its success stories is the Regional Seas Programme which brings together 120 countries and 14 UN agencies to deal with marine pollution and coastal degradation.

Besides collecting information regarding the environment, UNEP also disseminates environmental education. For this purpose, it organizes a number of conferences and workshops world-wide, and often in collaboration with other organizations, both national and international. A combined UNEP–UNESCO Environmental Educational Programme is in place. The UNEP also maintains a register of NGOs of the world, the number of which has increased tremendously in the 1970s and 1980s. By the late 1980s, about 7,000 NGOs were listed with the UNEP Environmental Liaison Centre.

It should be noted that the major role of UNEP is information-related. The actual environmental supervision or project impact evaluation is carried out by other international agencies. Such agencies include the United Nations Development Programme (UNDP), the World Bank and other international lending agencies.

The Brundtland Report, 1987

The report of the Brundtland Commission could be considered as the next benchmark event after the Stockholm Conference. This report, published under the title *Our Common Future* (WCED, 1987), has guided, to a great extent, our concept of environmental priorities and management.

The Commission which produced the report is the World Commission on Environment and Development. It was established by a 1983 UN General Assembly resolution which called for a global enquiry into the state of the world. The Commission included twenty-three commissioners from twenty-two countries. They had diverse backgrounds, and they came from countries which spanned the developed and the developing world and also the communist block. Gro Harlam Brundtland, who had been the Prime Minister of Norway, chaired the Commission.

The Commission held open deliberations and had participatory meetings in a number of countries at which people from various vocations could express their views on the environment. Such contributions illustrated very well how common and global human concern about environment had become in the second half of the twentieth century.

The Commission looked specifically into certain key areas concerning population, environment and development such as energy, industry, food security, human settlements and international relations. The final report was submitted in 1987 and included a detailed account of the state of the environment of the world.

However, the Brundtland Commission is best known for the identification of the environmental objective which they called *sustainable development*. Environment and development were seen as inseparable. This particular concept has since then significantly influenced the management of environment and highlighted the link between environment and development. The Brundtland Commission also highlighted a number of issues, of which the three following are of special importance:

- international co-operation is essential but very difficult to achieve;
- national environmental agencies should be strengthened;
- a world conference on environment should be held and a UN programme on sustainable development established.

Along with sustainable development these recommendations have guided environmental policies and management at the international level since the submission of the report.

Sustainable development

The Brundtland Commission defined sustainable development as meeting 'the needs of the present without compromising the ability of the future generations to meet their own needs'. The concept therefore recommends both economic growth and environmental protection. There is an apparent conflict between the two goals, but according to the Commission a reconciliation is possible, provided fundamental changes in managing the global economy can be arranged.

The concept of sustainable development has been widely accepted but its perception varies. It has been interpreted across the whole range, from being an extremely environment-friendly concept to an implicit business-as-usual statement. It is primarily a working concept which alerts people to the danger of uncontrolled growth for immediate gains, but some academics have tried to define it rigorously and thereby grown disenchanted with the concept. It is impossible to have economic development without any environmental impact. It is, however, possible to limit the impact, and in some cases, also to strengthen beneficial impacts such as the availability of clean drinking water as part of urban development. Sustainable development as a goal was endorsed by 178 countries at the 1992 world conference at Rio de Janeiro, and since then its implications have been used to define developmental and environmental strategies at various levels: global, national and local.

The following problems also need to be solved before sustainable development can become an operational reality:

• Financing sustainable development has been difficult. The Earth is a holistic unit, and sustainable development therefore involves all countries. Funding estimates and terms and conditions of sharing costs have become contentious issues between the developed and the developing countries.
• Our knowledge about the physical environment is limited and this makes it difficult to determine the carrying capacity of various ecosystems and the thresholds of physical degradation beyond which activities such as land erosion, accentuated flooding, groundwater contamination and so on begin.
• We need to know about the ability of the future generations to substitute technology for resources.
• Sustainable development and abject poverty cannot co-exist.
• A global political consensus does not exist at the level required for achieving sustainable development. Such consensus is necessary to control resource overuse and mitigate inequities.

Although sustainable development was widely accepted, its progress did not meet expectations. Writing about ten years after the Commission's report, Brundtland herself noted the lack of progress towards a new world order with burden-sharing, common perceptions and shared responsibilities. The lack of progress towards the generally accepted goal of sustainable development can be attributed to a low level of international co-operation and weak political arrangements. As its expression becomes more and more popular across a wide range of people (academics, journalists, bureaucrats, politicians), its objectives become more and more diffused and less attainable.

In the search for merging sustainability with development, perhaps it is wiser to constrain ourselves to a few simple guidelines. Some of these are already in practice but their continuation in spite of the lack of progress to date is essential:

- Any significant loss of natural resources or degradation of the physical environment should be prevented. A Hippocratic 'do no harm' policy should be accepted in face of future uncertainties.
- Knowledge regarding the physical environment should be extended on an emergency basis regarding both understanding of the operating natural processes and building of databases.
- Extreme poverty should be mitigated along with regional imbalances and inequities, and the quality of economic growth should be monitored.
- Efforts to build a global consensus on sustainable development should continue. This will be a slow and arduous process, and will require innovative adjustments and a new mindset necessary for individual nations to rise above their narrow short-term self-interest.

The Earth Summit, Rio de Janeiro, 3–14 June 1992

This conference is formally known as the United Nations Conference on Environment and Development (UNCED), but it has also been called the Earth Summit or Rio Conference. The conference developed out of several factors: the recommendations of the Brundtland and other global commissions; a number of international treaties and conventions such as the Montreal Protocol in 1987 or the Law of the Sea in 1982; and publications by the UNEP and the International Union for Conservation of Nature and Natural Resources (IUCN) dealing with future global strategies related to the environment.

As in Stockholm, a series of preparatory meetings were required to determine the agenda of the conference and to reconcile the differences in opinion among the developed and developing countries and China. The differences were in determining priority of issues, sources of funding to tackle environmental problems, and the question of economic development. Unlike Stockholm, this was also a political event attended by 178 governments including heads of states and important governmental ministers. It was a much biggerevent with parallel meetings involving environmental groups and technical people, including industrialists. A parallel NGO forum was held. In all more than 30,000 participants came to Rio de Janeiro, a reflection of the globalization of environmental awareness and involvement. The media coverage spread the news of the conference across the globe. Most governments submitted a state-of-the-environment report for their individual countries which were very useful data sources. Furthermore, in many countries preparation of the report involved participation of people outside the government, and as such helped in building a sense of co-operation and contact in environment-related issues.

All this heightened a sense of expectation which according to some people was not fulfilled, and controversies exist regarding the success of the outcome of the Earth Summit. The major positive effects of the UNCED may be summarized:

- a set of agreements were accepted by the participatory governments;
- the principle of sustainable development was accepted along with future global administrative arrangements for environmental management;
- the Conference brought together the global community on environment.

The UNCED agreements

The following agreements came out of the UNCED, although there have been delays in formal ratification of the declarations by certain countries. It has also been questioned whether the agreements are always reflected in national actions on environment:

- *Framework Convention on Climate Change*: a legally binding treaty to lower the emission of the greenhouse gases, particularly carbon dioxide;
- *Convention on Biological Diversity*: a legally binding treaty to prevent the loss of biological diversity, protection of species, and sharing of related research information and technology;
- *Rio Declaration*: a set of principles linking environment and development;
- *Agenda 21*: a large document regarding the action programme for sustainable development in the next century;
- *Forestry Principles*: a plan for forest protection and management; the accepted version was a reduced form of a comprehensive original draft.

The other topics which were discussed at length included the threat from desertification, the role of women in development and environment, and institutional arrangements for environmental management and funding. The financial resources and mechanisms were arguably the least achieved result of the UNCED.

The post-UNCED environmental scene

A series of changes evolved in the years after the Conference at Rio de Janeiro. Such changes were at least partially influenced by the Brundtland Report, the concept of sustainable development and agreements at Rio. Such changes affected environmental governance and objectives at various levels: global, national and local. New organizations were created and new groups became involved in environmental management. It is necessary to know about the current environmental actors and systems of management in order to comprehend the present state of environmental management. The biggest problem, however, remains the same: financing the proposals agreed to at the UNCED.

The United Nations organized a special session of the General Assembly in June 1997 in New York, popularly known as the *Earth Summit plus Five*. The meeting was attended by government representatives and environmental activists. The purpose of the meeting was to review and appraise the progress made in realizing the objectives set out at Rio de Janeiro in 1992. In many ways it was a disappointing meeting. The General Assembly concluded that in spite of the agreements at UNCED, the quality of the environment had continued to deteriorate. Globally pollution emissions including toxic substances and greenhouse gases have risen, although progress has been made in some countries to tackle this problem. Other problems highlighted at Rio (diminishing global biodiversity, transboundary movement of hazardous and radioactive wastes, acid rain, poverty-driven accelerated degradation of natural resources, unsustainable use of resources, inadequacy of water supply, etc.) continue to degrade the environmental quality.

Although promises made in Rio were reaffirmed in New York, it was evident that serious and tangible steps should be taken immediately in order to ameliorate the environmental conditions of the world. The lack of commitment from the developed nations was noticeable. Viable co-operation between the developed and the developing countries is essential for proper environmental management of the world. There is cause for apprehension following the failure at the 1997 international meeting in Kyoto (due mainly to the low-key approach of the United States) to strengthen the Convention on Climate Change, and thereby ameliorate global warming.

New developments at the global level

The United Nations General Assembly established several organizations to implement the recommendations at UNCED, and particularly regarding sustainable development and Agenda 21. The United Nations Commission on Sustainable Development (CSD) was formed in early 1993 to supervise progress in implementing Agenda 21, to oversee issues of sustainable development, and to co-ordinate issues of environment which are multisectoral and multicountry. Other agencies were established to co-ordinate matters pertaining to Agenda 21. The United Nations Development Programme (UNDP), the body associated with developmental work and assistance, has been building both technical and institutional capacities for carrying out the work specified in Agenda 21. This development is called Capacity 21. The objective of Capacity 21 is to make certain that individual countries are able to follow and administer the principle of sustainable development and the objectives of Agenda 21. Again, the UNDP has developed in-house techniques for evaluating all development projects which receive UNDP assistance, financial or technical, for possible environmental impacts. Similarly, the World Bank, a major donor agency, now requires all development projects to be assessed for possible environmental impacts. Over time, the Bank has become more attuned to environmental issues and has progressively integrated environment with development.

Global Environment Facility

Global Environment Facility (GEF) was designed as a mechanism for providing financial assistance to developing countries for dealing with specific environmental issues which affect the global environment. The support extends to the following areas:

- reducing global warming;
- reducing stratospheric ozone depletion;
- protecting biological diversity;
- protecting international waters.

Small amounts of funds could also be used to support other environment-related projects. GEF is also expected to act as a catalyst for focusing attention on global issues, mobilizing resources, and linking national and international efforts in its areas of concern. For example, GEF has supported the development of wind power in Costa Rica, the reduction of ozone-depleting substances in the Czech and Slovak Federal Republics, and the management of coastal wetlands in Ghana.

GEF is managed by three international agencies: the World Bank, UNEP and UNDP, with the World Bank in the chair. They have different responsibilities. The UNEP is more concerned with research and with selecting suitable project areas. The UNDP's primary role is in co-ordinating development with environment and organizing technical assistance through its network of offices in the developing countries. The World Bank administers the trust fund.

The pilot phase of the project ran for three years, 1991–4. The funding came from various sources such as a group of countries, including several developing countries, and the Multilateral Fund for Montreal Protocol which was set up to phase out ozone-depleting substances. GEF was restructured after 1994 and an expansion of its membership is expected. Currently, all countries which either have a UNDP programme in place or is able to borrow from the World Bank are eligible for GEF funding.

GEF has been criticized primarily for its modest amount of funding and for its loosely defined role in global environmental protection. GEF has not always met the high expectations it raised at its inception, but it is an example of a global effort towards environmental management. Its shortcomings highlighted at least two important requirements for success: better integration of national and international efforts, and the need to make lifestyle adjustments across the entire world.

National and local arrangements

Management and protection of the environment also require, besides a global consensus, careful governance at national and local levels. It is therefore important to motivate and establish efficient environmental organizations at these levels. Unfortunately, the capacity for environmental supervision varies from country to country and a Ministry of Environment or its equivalent, to oversee countrywide environmental issues, is often not a powerful ministry, at least in the developing countries.

The Environmental Protection Agency (EPA) of the United States, which started in 1970, set the model for a centralized organization to supervise and co-ordinate environmental matters. However, equivalent ministries set up by national governments may have two shortcomings. First, some environmental work may also be done (and for quite some time in the past) by other ministries which look after forests, water supply or biodiversity. This requires considerable co-ordination. Second, because the environment ministry is not necessarily considered to be powerful, it finds it difficult to control industrial pollution or to modify development projects. It is therefore necessary to modify the national administration's mindset about the priority of environmental issues. Understanding the domestic political economy is also essential to ensure higher priority for environmental concerns.

Similarly, a number of environmental issues have to be tackled at the local level, at least partially. Examples include air pollution in cities or sharing irrigation water. Environmental management not only has to be carried out at various levels but also needs to be co-ordinated among the three levels: global, national and local. Furthermore, the co-ordination should extend beyond the official governing bodies and involve the people who are directly affected. Often that requires working in collaboration with NGOs.

The non-governmental organizations

The importance of the contributions of the voluntary non-governmental organizations in preventing ecological degradation has been repeatedly demonstrated. Such groups occur at various levels, from grassroots to international, with varying interests and capabilities. Their existence at various levels, but with some linkages, is necessary to identify a local ecological problem, evaluate it technically, and draw the attention of the people and government to it. For example, at the grassroots level, the *Chipko* or the *Appiko* movement of India has prevented deforestation on steep slopes by outsiders armed with tree-felling contracts, by the simple and expeditious technique of hugging the trees when they are about to be felled. The movement has drawn regional and national attention to the problem of deforestation, soil erosion and increased flooding.

NGOs currently number in thousands and the total number of individual organizations has increased rapidly. A number of NGOs now have a global network, which adds to their strength. The work of the NGOs is recognized by a number of national governments, the UN institutions and the World Bank. However, the distribution of the NGOs remains variable across the world. For example, the movement is weak in China and North Africa. In certain developing countries, NGOs continue to be small, with a local base, and are under pressure. In some countries they are even considered hostile to the national government or powerful organizations. Their survival under such circumstances becomes difficult and depends on the courage of individuals.

The NGO movement is now sufficiently strong enough that in most developing countries a project will not be implemented without being examined and publicized by at least one NGO. In that sense, NGOs behave as local guardians of the environment with or without governmental co-operation. A number of NGOs currently have progressed beyond building a community movement. These include economists, scientists and technical personnel apart from grassroots leaders, and are capable of producing a sophisticated evaluation of the possible environmental impacts of a proposed development project. NGOs publish such studies as well as general reviews of the country-wide state of the environment. In spite of many difficulties, NGOs have become an important part of the environmental management scene.

Industries and environmental management

Traditionally, industry has been associated with various acts of pollution which are controlled by government regulations. Currently, industry itself is moving progressively towards reducing harmful emissions. This is done through innovative attempts to minimize waste material and pollution. An example is the strict accounting of the material used so that material that degrades the environment is dealt with inside the plant instead of being released in emissions. A new term, *industrial ecology*, has been coined to describe this approach which requires that minimal damage to the environment is caused by industrial products. This approach is still at the formative stage but at least it is a positive sign that in the near future industrial establishments may themselves be part of active management of the environment. There is also the need for technology transfer for this to happen world-wide, and particularly in the developing countries.

These attempts are helped by developments such as marks of approval provided by governments, for example, the Green Seal of the United States or the Blauer Engel of Germany. The International Standards Association's ISO 9000 (whose new development is ISO 14000) is another publicized guarantee of reasonable standards of quality in production. It is clear that in the industrial world an attitudinal change is in progress (Schmidheiny, 1992), and the future might see an enlightened industry committed to maintaining the quality of the environment. Such changes should also affect the industry of the developing world.

The Brundtland Commission expressed the dichotomy between the physical and the political and economic worlds as 'The earth is one, but the world is not'. The success of future environmental management depends on the unification of the political and economic world.

Key ideas

1 The Stockholm Conference of 1972 internationalized the environmental movement and established the linkage between environment and development.
2 The report of the World Commission on Environment and Development (also known as the Brundtland Report) has largely guided our concept of environmental priorities and management since its publication in 1987, mainly through the concept of sustainable development.
3 The concept of sustainable development, as proposed in the Brundtland Report, requires meeting the needs of the present without jeopardising the ability of future generations to meet their own needs.
4 The United Nations Conference on Environment and Development (UNCED) was held at Rio de Janeiro in 1992.
5 The UNCED led to a number of agreements including the Framework

Convention on Climate Change, Convention on Biological Diversity and Agenda 21.
6 A number of post-UNCED developments such as the Global Environment Facility have become part of environmental governance.
7 Proper environmental governance requires co-operation at all three levels: global, national, local; new mindsets; and more equitable economic and political arrangements.
8 The lack of funding and political will remain the biggest obstacles to the sustainable development of the world.

9

Environmental problems and the Third World development

A brief recapitulation

Two sets of observations run through the text of this book. First, many acts of development in the Third World have left a series of environmental problems; and second, the environment operates as an integrated system, so any type of modification, even if it is local in nature, may start a chain of events resulting in multifarious effects, regional in scale. Examples of this can be seen in some of the cases discussed: the deforestation of the tropical rainforest; construction of dams across rivers; groundwater depletion to meet urban demands. Such effects could even be global in some cases, as exemplified by the stratospheric ozone depletion and global warming.

It should be pointed out that different types of environmental degradation may arrive with different kinds of economic activities. Air pollution, for example, occurs where a large amount of fuel combustion and industrial activity take place. In that sense, the chapters in this book are arranged progressively in order of more complicated and extensive environmental degradation. Most Third World countries suffer from the effects of resource extraction or the expansion of agriculture to marginal areas. Only the more technologically advanced and populated ones pollute the water and air to a high degree. The non-uniform availability of information must also be noted. A high frequency of examples from a country not only indicates that cases of environmental degradation are common, but also that such ecological modifications are known, perhaps are being monitored, or even that some preventative measures are being taken. Absence of examples does not necessarily indicate good environmental management. One essential step towards preventing environmental degradation is awareness and involvement at various levels. National governments, provincial and local administrations, academics, the legal profession, the press, the business community and the local inhabitants all have to live in the area of ecological deprivation. Over the last three decades, the countries of the Third World have become increasingly aware of the dangers inherent in unchecked and large-scale development projects. It is now even politically acceptable to consider the environment.

The machinery for environmental protection

Development programmes which take into consideration the possible negative environmental impacts and take steps to reduce or prevent them are thus desirable for the Third World. This is being progressively formalized, and in many countries a large-scale development project would require an environmental impact assessment (EIA) or equivalent before the project could begin. This could be a national requirement. If the project is planned with financial or technical assistance from international donor organizations such as the World Bank or the United National Development Programme, the donor agencies also evaluate the project for possible environmental impacts.

Mechanisms for protecting the environment at the global level are developing, although not as rapidly or as efficiently as many would like. These are:

- organizations such as the United Nations Commission on Sustainable Development or the UNEP;
- the in-house arrangements of the international donor agencies such as the World Bank or the UNDP to investigate proposed development projects for possible environmental impacts;
- a range of international agreements such as the Montreal Protocol against stratospheric ozone depletion or the Law of the Sea.

Not all these agreements are strongly adhered to or properly enforced, and this is particularly true regarding the management of the tropical rainforests. However, a beginning has been made, and treaties such as the framework Convention on Climate Change and the Convention on Biological Diversity agreed to at the UNCED in 1992 are good examples. Probably the most important development has been the change, although partial, in the mindset of people such as administrators or industrialists. The two international conventions of 1972 and 1992, the Brundtland Report with the working concept of sustainable development, and the scientific explanations of stratospheric ozone depletion and global warming have done that. At the national level, this awareness has resulted in the establishment of governmental machinery such as environmental ministries and an acceptance of the requirement for evaluating the environmental impact of proposed large-scale development projects such as large dams or roads through a rainforest. However, a great need exists to collect basic information regarding both physical and socio-economic environments at the national level for many developing countries. This type of baseline data is imperative for correct environmental management.

Apart from the need for technical knowledge and long-term data on environmental conditions, the developing countries need strict legislation to deal with environmental degradation such as water and air pollution. Such legislation already exists in some Third World countries although their implementation is not necessarily common. Furthermore, countries should have government agencies with money, clout, and the technical expertise to evaluate, monitor and, if necessary, modify or restrain projects for development. This is not always forthcoming, even in the developed countries. The importance of the contributions of the voluntary non-governmental organizations (NGOs) in preventing ecological degradation has been repeatedly demonstrated. Such groups occur at various levels, from grassroots to international, with different interests and capabilities. Their existence at different levels, but with some linkages, is necessary to identify a local environmental problem, technically evaluate it, and draw the attention of the people and the government to it. Over the last few years, the industrial establishment has also become environment conscious, as discussed in Chapter 8. Environmental evaluation and management is currently the combined effort of a number of groups with different expertise and interests.

There are, however, elements of chance and of local politics in the prevention of possible environmental deterioration associated with the setting up of an industry or of a large-scale hydroelectric project, as illustrated by the case of the Silent Valley.

Both Silent Valley and the *Chipko* movement illustrate that for successful environmental protection a combination of environmental actors working at various levels against development projects which degrade the environment is crucial. It is also more profitable to work in some kind of collaboration with government agencies rather than in total opposition. Both the post-UNCED change in the mindset of administrators at various levels (even if partial) and good environmental science help in building a communication bridge. What is required is not a total confrontation but an enlightened discussion involving all sides, based on good science, knowledge of the area and the goal of sustainable development. There is a long way to go, but it is now almost impossible anywhere in the world to carry out a project with considerable environmental

Case study M

Silent Valley

Silent Valley is the narrow valley of the Kunthi River in the state of Kerala in south-western India, at an elevation between 1,000 m and 2,400 m. Apart from its scenic appearance, the valley has 8,950 ha of rainforest with valuable rare plants and animals, including the lion-tailed macaque, a threatened primate. The narrow gorge at the lower end of the valley has been considered several times over a number of years as a possible site for generating hydroelectricity. In 1973, the then state government of Kerala wanted to dam the gorge, creating a reservoir upstream, which would have submerged a large part of the pristine rainforest, to generate power. The project, however, was delayed until 1976.

By that time the Silent Valley project had drawn the attention of several prominent people with an interest in conservation. They included, among others, officers in the central government in New Delhi, and a government ecological task force. The task force report queried the cost estimates of the project, was unhappy about the submergence of the forest, and wondered whether the proposed generation of only 120 megawatts of power justified the destruction of such a unique environment. About that time a group of school and college teachers and scientists of Kerala started a protest against the project on the grounds of environmental despoliation.

The project by this time was being attacked by various Indian and international conservation groups, resulting in attempts by the state government to rush the project through. The state assembly passed a unanimous resolution for the speedy implementation of the hydroelectric project. The implementation of the project was delayed by scientific controversy, several court orders, and the fact that approval was needed from the central government. While all these were being sorted out, Indira Gandhi returned to the premiership in early 1980 and, being sympathetic to conservationist issues, referred the project to a new scientific committee which in 1983 came out in favour of conservation.

Silent Valley has become a landmark in the ecological movements, where a Third World group of conservationists prevented the state government from destroying a valuable rainforest. It is, however, salutary

Case study M (*continued*)

to note that the success of the movement was due to: the fortuitous coming together of several conservationist groups at various levels, local, national, international; the willingness of some scientists to examine the issue and take a stand on their findings; the publicity generated; and the fact that, when the time for the final decision arrived, the prime minister happened to be a person interested in environmental protection (D'Monte, 1985). There is an element of chance in the protection of the environment.

impact without prior detailed assessment and public knowledge. It is only possible under extremely autocratic political situations. In the fifteen years or so since Silent Valley, a number of projects with high potential for environmental damage have received considerable publicity during the planning and inception stages. Some, like the Nam Choan Dam in Thailand, have even been abandoned.

The model of large-scale development may not be appropriate for everybody and technology has to meet the scale of the demand and the environment. Small is sometimes, even if not always, beautiful. Water management in the irrigated rice areas of Bali is carried out by the *subak*, an irrigation association which has been in existence for centuries. The members of the association are farmers who grow irrigated rice; under an elected head they are collectively responsible for monitoring the irrigation conduits and the infrastructure and for managing the crop pattern. The organization is an independent entity but works in coordination with the infrastructure of the local government. The fields of Bali have thus been efficiently supplied with water for hundreds of years without the need for a large-scale engineering project.

The future trends

There has been a heightening of awareness towards environmental protection in the developing countries. There is a widespread acceptance of the need to maintain an environment of high quality. At the same time development and protection of the environment are readily accepted as interrelated, although one or both of them might have to be done on a crisis footing. As the countries of the Third World become industrialized, they inherit problems similar to those of the developed countries, that at times are added on to the pre-existing problems of dense population, inequity and excessive extraction of non-renewable resources. On occasion, however, a diffusion of technology and management from the developed countries regarding environmental problem-solving does occur.

A common problem for a number of the Third World countries is the pressure on them to service their international debt. As a result, some countries tend to over-exploit their natural resources. Two examples of this situation, discussed by Myers (1986), are cattle ranching and the easing of logging restrictions in the tropical rainforest. The burden of the international debt is also behind the use of farmlands for non-food crops for export commodities, thereby pushing poor farmers on to the marginal areas. The conflict between development and environmental protection, or between the interests of the developed and developing countries, is due to overlooking the fact that there is only one Earth. If rainforests are to be preserved for the sake of the world, it is necessary to provide technical and financial help to the countries that still have the rainforests, so that development is possible without uncontrolled deforestation. This is a general principle extendable to other areas of environmental management.

Although many of the proponents of sustainable development and those who followed the UNCED at Rio de Janeiro in 1992 are at least partially disappointed at the slow pace of environmental co-operation and management, the world has come a long way in developing mechanisms and agreements in this area. Problems still exist in the transfer of technology and funding from the developed to the developing countries in sufficient amounts to make a significant impact on the environment at all levels. The world pattern of trade and political arrangements also do not favour the developing world. Such problems need to be solved for the sake of a better physical world where inequity is reduced across both geographical and social space.

In March 1985, the Supreme Court of India, petitioned by the citizens of the Doon Valley in the fragile environment of the Kumaon Himalayas, declared that a large number of limestone quarries in the valley should be closed on grounds of environmental degradation. According to the judgment this has to be done in the interest of protecting and safeguarding the right of the people to live in a healthy environment with minimal disturbance of the ecological balance. This historic judgment indicates that uncontrolled development cannot take precedence over environmental protection. Growth of per capita income, equity of resources and sustainability of the environment are simultaneously required in the Third World.

Key ideas

1 Development is related to the improvement of both economic and environmental conditions.
2 Development programmes should take into consideration the possible degrading environmental impacts and take steps to avoid or reduce such problems.
3 Environmental protection is possible only when backed by legislation, efficient government machinery, and public awareness and participation.
4 There is an element of chance in the protection of the environment.
5 Considerable awareness has developed towards environmental protection in the Third World.
6 A hierarchical arrangement for environmental management is developing which operates at various levels: global, national and local.
7 There is only one Earth; we all share its resources.

Review questions, references and further reading

Chapter 1

References and further reading

Carson, R. (1962) *Silent Spring*, Harmondsworth, Penguin.
Marsh, G.P. (1898) *The Earth as Modified by Human Agencies: A Last Revision of 'Man and Nature'*, New York, Charles Scribner & Sons.
World Resources Institute (1994) *World Resources 1994–95: People and the Environment*, New York, Oxford University Press.

Chapter 2

Review questions

1 Where is relatively undisturbed natural vegetation found in the Third World? Are these areas under the threat of deforestation? If so, what type of land use will replace the natural vegetation in such areas?
2 Why is the issue of fuelwood so important in the developing countries? What kind of hardship is caused by a scarcity of fuelwood? Is there any viable alternative?
3 The degradation of the environment as a result of deforestation is felt at various levels: local, regional, global. What are the effects of deforestation?
4 What is the role of the First World organizations in the deforestation of the Third World?
5 What steps can you suggest for preserving the tropical rainforests? Why should we preserve the tropical rainforests?

References and further reading

Agarwal, A. and Narain, S. (eds) (1986) *The State of India's Environment 1984–85:, The Second Citizens' Report*, New Delhi, Centre for Science and Environment.

Anderson, J.M. and Spencer, T. (1991) *Carbon, Nutrient and Water Balances of Tropical Rain Forest Ecosystems Subject to Disturbance: Management Implications and Research Proposals*, MAB Digest 7, Paris, UNESCO.

Jackson, P. (1983) 'The tragedy of our tropical rainforests', *Ambio*, 12, 252–4.

Myers. N. (1986) 'Economics and ecology in the international arena: the phenomenon of linked linkages', *Ambio*, 15, 296–300.

Repetto, R. (1990) 'Deforestation in the tropics', *Scientific American*, 262(4), 18–24.

Salati, E., Dourojeanni, M.J., Novaes, F.C., De Oliveira, A.N., Perritt, R.W., Schubart, H.O.R. and Umana, J.C. (1990) 'Amazonia', in B.L. Turner II, W.C. Clark, R.W. Kates, J.F. Richards, J.T. Mathews and W.B. Meyer (eds), *The Earth as Transformed by Human Action*, Cambridge, Cambridge University Press, 479–93.

World Resources Institute (WRI) (1994) *World Resources 1994–95: People and the Environment*, New York, Oxford University Press.

Chapter 3

Review questions

1   What are the factors that cause pressure on land in the developing countries?

2   Where are the marginal lands located in the tropics and subtropics?

3   How does the Green Revolution work? Does it cause any environmental problems? Has the Green Revolution been successful?

4   What is the proper way to bring irrigated water to dry lands?

5   What is desertification and where does it happen? Is there a solution?

6   What are the effects of deforestation and cultivation on steep slopes? Are the effects all local? What parts of the tropics and subtropics are affected by cultivation on steep slopes?

References and further reading

Eckholm, E.P. (1978) *Losing Ground*, Oxford, Pergamon.

Kendall, H.W. and Pimentel, D. (1994) 'Constraints on the expansion of the global food supply', *Ambio*, 23, 198–205.

Olsson, L. (1993) 'On the cause of famine: drought, desertification and market failure in the Sudan', *Ambio*, 22, 395–403.

Stáhl, M. (1993) 'Land degradation in East Africa', *Ambio*, 22, 505–8.

Chapter 4

Review questions

1 What are the different types of water use? How does demand for water change with improving economic conditions?
2 How much water do you use in 24 hours? Keep a record of your use of water for several days in order to determine this.
3 Why is safe drinking water so important? How prevalent is the supply of potable water in the Third World? What is the indirect benefit of bringing close to the inhabitants of the Third World?
4 Select several cities. Determine their sources of water supply. Does any of your cities have a shortage of water? of required quality?
5 How do rivers purify themselves of organic wastes? Select a large river in the tropics or subtropics. How many cities use it both for supply of drinking water and disposal of waste water?
6 What are the major sources of water pollution?
7 Select a major project on a river within 30° of the equator. Tabulate the benefits and problems arising out of the project. Are there any possible solutions to the problems? Do you approve of the project?

References and further reading

Biswas, A.K. (1978) 'Water development and environment', in B.N. Lohani and N.C. Thanh (eds), *Water Pollution Control in Developing Countries*, vol. 2, Bangkok, Asian Institute of Technology.

Dunne, T. and Leopold, L.B. (1978) *Water in Environmental Planning*, San Francisco, W.H. Freeman.

Hammerton, D, (1972) 'The Nile River: a case history', in R.T. Oglesby, C.A. Carlson and J.A. McCann (eds), *River Ecology and Man*, New York, Academic Press.

World Bank (1992) *World Development Report 1992: Development and Environment*, Oxford, Oxford University Press.

World Resources Institute (1996) *World Resources 1996–97: The Urban Environment*, New York, Oxford University Press.

Chapter 5

Review questions

1 What are the major air pollutants? List their sources and effects.
2 Why is indoor air pollution common in the developing countries?
3 Select a developing country with considerable mining activities. How much air

pollution would you expect near the mines? What are your suggestions for dealing with such problems?

4 Air in a number of Third World cities is highly polluted. Choose one such city and assess the level and sources of air pollution. Can you suggest any remedial measures?

5 What are air pollution standards?

6 How would you ensure that a Bhopal-type disaster will not happen again in the Third World?

References and further reading

Agarwal, A. and Narain, S. (eds) (1986) *The State of India's Environment 1984–85: The Second Citizens' Report*, New Delhi, Centre for Science and Environment.

Masters, G.M. (1991) *Introduction to Environmental Engineering and Science*, Englewood Cliffs, N.J., Prentice-Hall.

Miller, G.T. Jr (1986) *Environmental Science: An Introduction*, Belmont, N.Y., Wadsworth.·

Chapter 6

Review questions

1 In what way does the physical environment of a city differ from that of the countryside?

2 How does the building of a city modify the local climate? What effect does it have on the well-being of the citizens?

3 Explain why urbanization is followed by increased flooding. What steps would you suggest to ameliorate such flooding?

4 How do the cities in developing countries bring people and pollution together?

5 Rapid urbanization is occurring right now in the Third World. What is it doing to the environment?

References and further reading

Gupta, A. (1984) 'Urban hydrology and sedimentation in the humid tropics', in J.E. Costa and P.J. Fleisher (eds), *Developments and Applications of Geomorphology*, Heidelberg, Springer-Verlag.

Landsberg, H.E. (1981) 'City climate', in H.E. Landsberg (ed.), *General Climatology 3*, World Survey of Climatology, vol. 3, Amsterdam, Elsevier.

Rau, J.L. and Nutalya, P. (1982) 'Geomorphology and land subsidence in Bangkok, Thailand', in R.G. Craig and J.L. Craft (eds), *Applied Geomorphology*, London, Allen and Unwin.

United Nations (1995) *World Urbanization Prospects: The 1994 Revision, Estimates and Projections of Urban and Rural Population and of Urban Agglomerations*, UN Publication ST/ESA/SER.A/150, New York, United Nations.

Wolman, M.G. (1967) 'A cycle of sedimentation and erosion in urban river channels', *Geografiska Annaler*, 49A, 385–95.

World Resources Institute (1996) *World Resources 1996–97: The Urban Environment*, New York, Oxford University Press.

Chapter 7

Review questions

1 Explain the depletion of the stratospheric ozone layer. What are the possible effects?
2 What are the uses of chlorofluorocarbons (CFCs)?
3 CFCs are currently being phased out. How did it happen?
4 What is the ozone hole over the Antarctic? Are there ozone holes elsewhere?
5 What is global warming? Is global warming always detrimental?
6 List the effects of global warming. Now determine which ones on your list are relatively uncertain.
7 What would be the effect of global warming in the tropics and subtropics?
8 What can we do to slow down global warming?

References and further reading

Barron, E.J. (1995) 'Researchers assess projections of climate change', *Earth on Space*, 8(1), 4–5.

Intergovernmental Panel on Climate Change (1994) *Radiative Forcing of Climate Change: The 1994 Report of the Scientific Assessment Working Group on IPCC. Summary for Policymakers*, IPCC.

Jones, P.D. and Wrigley, T.M.L. (1990) 'Global warming trends', *Scientific American*, 263(2), 66–73.

Molina, M.J. and Rowland, F.S. (1974) Stratospheric sink for chlorofluoro-methanes: chlorine-atom catalysed destruction of ozone', *Nature*, 249, 810–12.

Rosemarin, A. (1990) 'Some background on CFCs', *Ambio*, 19, 280.

Rowland, F.S. (1990) 'Stratospheric ozone depletion by chlorofluorocarbons', *Ambio*, 19, 281–92.

Chapter 8

Review questions

1 Why are global arrangements and co-operation necessary to protect the environment?
2 What did the Stockholm Conference accomplish?
3 What is sustainable development? Is it feasible?

4 Was the UNCED meeting successful? Did it change anything?
5 Why do we need a hierarchical structure of global governance at various levels?
6 The Brundtland Commission said that the Earth is one, but the world is not. Why?

References and further reading

Schmidheiny, S. with the Business Council for Sustainable Development (1992) *Changing Course: A Global Business Perspective on Development and the Environment*, Cambridge, Mass., MIT Press.
World Commission on Environment and Development (1987) *Our Common Future*, Oxford, Oxford University Press.

Chapter 9

Review questions

1 Why should development of an area take into consideration the environmental conditions?
2 What machinery do we need to protect the environment in the Third World? Do we have it?
3 How would you evaluate the future trends towards environmental protection in the developing countries?

References and further reading

D'Monte, D. (1985) *Temples or Tombs*, New Delhi, Centre for Science and Environment.
Myers, N. (1986) 'Economics and ecology in the international arena: the phenomenon of linked linkages', *Ambio*, 15, 296–300.
World Resources Institute (1994) *World Resources 1994–95: People and the Environment*, New York, Oxford University Press.

Index

Milton Keynes UK
Ingram Content Group UK Ltd.
UKHW040013071024
449327UK00011B/208